SMALL FARM SUCCESS AUSTRALIA

HOW TO MAKE A LIFE AND A LIVING ON THE LAND

ANDREW CAMPBELL

ANNA FEATHERSTONE

CAPEABLE PUBLISHING

The information in this book is the author's and interviewees' opinions and of a very general nature only. Always seek appropriate advice from qualified experts (accountants, agronomists, veterinarians, officials etc.) about your individual circumstances. Readers should not rely on the general information given in this book as a substitute for professional advice. The author, publishers and interviewees cannot accept responsibility for any losses, damages or adverse effects that may result from the use of information contained in this book.

Published by Capeable Publishing

First published in Australia in 2018

To order directly, please order online at www.smallfarmsuccess.com.au

ISBN: 978-0-9807475-1-5

Cover design: Gayna Murphy, Mubu Design

Cover image: nikkiphoto

Interior: Vellum

Printed by: Ingram Spark

A catalogue record for this book is available from the National Library of Australia

Dedication
To Fran, forever cooking up kindness and fuelling everyone around her with love.

"Why do farmers farm, given their economic adversities on top of the many frustrations and difficulties normal to farming? And always the answer is: "Love. They must do it for love." Farmers farm for the love of farming. They love to watch and nurture the growth of plants. They love to live in the presence of animals. They love to work outdoors. They love the weather, maybe even when it is making them miserable. They love to live where they work and to work where they live. If the scale of their farming is small enough, they like to work in the company of their children and with the help of their children. They love the measure of independence that farm life can still provide. I have an idea that a lot of farmers have gone to a lot of trouble merely to be self-employed to live at least a part of their lives without a boss."

— WENDELL BERRY, BRINGING IT TO THE TABLE:
ON FARMING AND FOOD

CONTENTS

INTRODUCTION

You could be reading a book right now about how to become an Instagram star, a tech entrepreneur or a top sales executive. You could be reading a book about how to invest for retirement, how to style your lounge room, or how to organise your cupboards. Instead, you're sitting here reading a book about farming. Why is that?

It's because a tiny seed has set root in your soul – and you've decided to water it.

I wasn't born a farmer, and growing up I lived at least a thousand suburban streets and a highway away from the nearest one. The closest I ever got to going bush was a trip every few years to drought-cracked Bakers Swamp, near Wellington, NSW. Here, family friends were building a mud-brick house while farming Corriedale sheep and a particularly vicious donkey. I never ate broccoli in my real life, but in their vegie patch I'd snap florets straight from the plant and eat them raw. Everything was fresher out there, everything tasted better out there, and even as a child I knew that I too was better out there. It was the space, the animals, the air – the freedom to explore, the grounding, the connection to earth. But the city and the beach were my family heritage, and air-conditioned office towers loomed well into my thirties.

Becoming a parent with my husband Andrew changed that. Raising the kids in the confines of ever-more expensive suburbia changed that.

Breathing in the negative aspects of consumerism, terrorism and mean-ingless work changed that. Then Andrew got sick.

"We've always talked about living on a farm," we agreed. "If not now, when?"

And so the adventure and healing began.

Nearly 4 000 days of farming later, 36 hectares of land has given us the best, most challenging, most uplifting, most energising, most difficult time of our lives, and we'd do it all again tomorrow...although we'd do it slightly differently. That is our motivation for writing this book – to bring together the wisdom and learnings of farmers from across Australia to help you on your way, so you won't make nearly as many mistakes as we did. It's also to give you inspiration, so you can see what is possible, so you can begin planning, acting on, or enhancing your current agricul-tural adventure.

Over the last decade, our family has made a life and a living from our award-winning small farm. Our produce has included organic turmeric, honey (from Italian and native Australian stingless bees), miniature Galloway cattle and rosellas (wild hibiscus). Our value added produce – sold at farmers' markets and online – ranged from herbal teas to honey orange mustard, habanero honey to organic beeswax lip and skin balms. We've raised goats, multiple breeds of sheep, chickens and alpacas, and two horses came along for the ride. We've given thousands of people experiences through farmstays, tours and courses. Mostly, though, we've given ourselves and our children a priceless experience. It's been a life filled with joy and struggle, practicalities and passion. It's been a life of both independence and community. It's been a life well worth living.

And living is what it is all about. Each of us only has so many decades of it. The first two decades are swallowed up by childhood, and the next one or two fly by. Then the next...and the next. Time keeps slip-ping away. It's great you're reading this book now to help you decide if you might spend at least a few of those precious years getting your hands dirty and giving your heart to the highs, the lows and the inestimable joys of farming.

Whether you're a student, hobby farmer, someone wanting to earn (or make a better!) living from farming or perhaps a backyard dreamer, I hope this book and the generous tips given by other farmers will help you on your journey. *Anna*

Throughout this book you will find short case studies on a variety of different farmers. These spotlights are just that, a look at a specific time in the lives of these people and their farming operations. Depending on when you are reading this book, these farmers circumstances may well have changed and they might be up to all sorts of different things. We thank them for their generosity in sharing their stories so you can learn from them.

PART I

1

WHY FARMING?

So much in modern culture distracts us from real life. There are pixels to play with, dramas to binge watch and sports to be sweated. There are careers to be plotted, investments to be planned and dinner-time takeaway options to be considered and ordered. There are friends' holiday snaps to like on Facebook, the lure of the next big sale, and credit cards packed thick in our wallets pretending to be money. It's all so luxurious and simple, so safe and repetitive...until you wake up.

Until you wake up and realise you're no longer fulfilled. And that not only are you not fulfilled, but you're frustrated and churning. You're no longer the child who dreamed, but a mortgaged, suited and looted skeleton of your former self. Even with all the bells and whistles of a 'real' job or degree, or trade certificate or fenced home in the 'burbs, you know there's a whole lot more to life than this.

Perhaps you want to run away from it all. Or perhaps, like a thoroughbred sniffing the air, you can sense a world of adventure and freedom awaiting you outside the city limits. Perhaps you want to gallop to a new life as fast as you can.

It's not every day someone decides to become a farmer, and of the more than twenty four million people living in Australia, only about 300 000 work in agriculture. More people work in the business of selling food and fibre than in growing it. But a lot of us dream of becoming a farmer, and what follow are the key reasons why.

To escape the rat race

If you feel suffocated by the 9 to 5 routine, surrounded by walls and framed by cubicles, counters and self-imposed ceilings, you're probably also slowly stewing your soul during peak hour, staff meetings and while standing in line in shopping centres.

While your unique fingerprints uniformly tap on mass-produced keyboards or swing tools or grasp the steering wheel as you drive the kids to sport, you're finding it harder and harder to bite your tongue when colleagues, customers, neighbours and bosses annoy you. Even when your job is a joy and the pay packet appears in your account regularly, you still know something's lacking.

What's not right? Perhaps it's the presence of buildings but absence of a horizon. Maybe the filtered hum of air-conditioning never flows over you like a real breeze carrying scents of eucalypt, nectar and blossom. Or maybe it's just that you're cut off from the deep connection to earth, animals and seasons that the human species needs to feel grounded.

If you're sick of keeping up with the Joneses and of monotonous weeks doing a job you no longer love, you're probably bumper to bumper with frustration. Country living with its free flowing (though often potholed!) roads might just be the balm you're looking for.

To raise children in the country

It's no wonder some parents feel that a country upbringing is beneficial for kids, because the bulk of city childhoods now look a little like this: kids on concrete, kids on screens, kids in the backseat constantly being chauffeured to distractions.

Kids on farms: well, you don't see them that much! But the branches and trunks of trees do; the slippery logs across creeks do; the flickering stars and dancing campfires do. So do the butterflies, the dew, the furry beasts and the grass upon which kids bound and cartwheel as they set off on their next adventure.

Though cities have enormous entertainment and organised activity offerings, farms give children the space they need to grow, explore and learn real life lessons. Children come alive when allowed to contribute, when challenged and when given responsibilities within their reach, and farms provide a daily dose of all three. Farm kids learn to roam and explore, and to problem solve. Daily, they have opportunities to interact with nature and are exposed to skills that help them grow into capable and caring human beings with a connection to the earth.

If you'd prefer to be begging the kids to come inside for dinner, rather than begging them to get off their devices, farming might turn out to be great for your family. It certainly provided our kids with an adventurous childhood while helping them develop all sorts of capabilities and life skills.

Family heritage

Did you grow up on a farm? Or did you visit extended family on theirs? Are you thinking of returning out of a sense of duty to ageing relatives, or maybe you never escaped the magnetism of the land? Perhaps, having experienced the city, you just want to give your children the same kind of upbringing you enjoyed. We often return to what we know, and if this is the case, your time away from the farm will hopefully have endowed you with great new ideas and a strong sense of purpose.

It's your vocation

You've studied horticulture, agriculture or animal science. You've listened to lectures about aquaculture, viticulture or forestry. You've been to beekeeping, permaculture and dairying courses. Now you're burning to put all that knowledge into action. You've already invested in your education; now it's time to get started on your calling.

For therapy and relaxation

According to Paul Clifford of Elders Real Estate in Victor Harbor, South Australia, many high-powered professionals buying into the region see a country property as therapy from their stress-loaded occupations. The mechanics of farming, combined with the serenity of it, gives them the ability to switch off and relax. Simple acts like mowing, weed pulling, fence fixing and feeding animals offer all the benefits of fancy exercise, mindfulness and relaxation, and can lower the stress levels that rise when you're in the hothouse of the office.

To live sustainably and self-sufficiently

A deep contentment springs from knowing you can supply your own food, water and energy. Even though cities offer community gardens and you can grow a reasonable amount of food in a suburban backyard, if you have a deep desire to be more self-sufficient, there's nothing like a farm to achieve these goals. Harnessing sun, wind and water, combined with growing your own food and getting involved with the local community, radically decreases your reliance on multinational companies and reduces some of the major expenses households face. The ability on a farm to recycle, compost, reuse and repair could enable you to drop a shoe size or two in terms of your carbon footprint!

To be your own boss

Tired of being trapped in the system, of always having to answer up the line? Want to back yourself and your business skills to run your own race? Though nature is the real boss on any farm, you'll feel alive working for her.

One of the key benefits of being your own boss is that you can make and quickly implement decisions to improve your business. No more going through a zillion meetings to get approval, no more having your best ideas end up in the waste basket. Being your own boss means you can just get out there and do it. This flexibility allows you to envision, innovate, experiment and implement – it also means you need to be prepared to make changes after you make mistakes. In any job, you're accountable to your boss, and in this case, that boss will be you.

For fitness and health

Some people will treechange after a health scare. A country move helps you and your family get away from plastics, pollution and the stress of city life; it allows you to surround yourself with the healing benefits of fresh air and fresh food.

Farming also requires lots of physicality, from walking, right through to digging, shifting, lifting, tugging, pulling and hauling. That's why one of the first savings you'll make on a farm is that you'll no longer need a gym membership! Agriculture's all-natural *farm boot camp* provides the ultimate workout for young and old. In addition, if you're growing your own food, you have access to fresh food, and your family will be much less likely to eat processed junk.

To work outdoors

There's a reason why we gaze out to the horizon, why we're mesmerised by sunrises, why we can't help but bend to smell a flower. There's a reason why wide, open skies makes us feel better than skyscraper-shaded streets, why breathing air filtered through trees beats air emitted from motors and why immersing ourselves in nature enriches us more than immersing ourselves in another marathon screen session. Being outdoors is enlivening.

You end up appreciating everything more when you work outdoors: cool early starts to avoid the heat of the day, hot showers after cold rain storms, the colours of spring and the slowing weeds of autumn. Perhaps you're thinking about farming to escape the claustrophobia of the city, or just to immerse yourself further in the riches of the planet.

To learn new skills

Someone once said to me that "if you do the same thing for long enough, you'll end up hating it." Lifelong learning, when you continually challenge yourself to learn new skills and stretch yourself to achieve mastery, is motivating and rewarding. If you've been stuck in an office and want to learn how to work with your hands, if you're a fabulous mechanic and wonder if you'd be able to tweak those skills to maintain farm machinery, if you've been a teacher and want to swap roles and be the student, a farm and all it entails could re-educate and reinvigorate you.

To save money

What you pay for a house in a capital city can buy you acres of land in a rural area, so you can downsize your mortgage while upsizing your life and land.

To make money

You know you want out of the city and your current workplace, but perhaps you're struggling to find suitable work in a regional area so you can make the transition. Part or full-time farming might be the answer to helping you fund and live a rural lifestyle.

The old joke goes, "How do you make a small fortune farming? Start with a big one." But more and more farmers are showing you can make a living from the land. You might have dreams of harvesting big bucks from unique crops and value added services, or maybe you just need enough income to support a simpler life. Whatever your goal, it will entail hard work because although money CAN be made in farming, it's not a get rich quick scheme and you won't be living the life of a Russian oligarch anytime soon.

For tax breaks and investments

Perhaps you're looking to reduce tax paid on other income streams, or are seeking an alternative investment for your super fund. Large acreages can offer tax advantages, as well as depreciation on plant and equipment, and tax deductions on improvements. Current GST rules can also be of advantage to farmers in that although GST is not currently charged on food, you can claim GST back on its production costs, for example on packaging and labelling. Primary producers might also qualify for diesel fuel subsidies, as well as be in the running for land and business improvement grants.

Love of animals

Fascinated by long-necked alpacas? Love your dog or cat but also have room in your heart for cheeky goats? Has owning a horse been a childhood dream that's ready for reality? Whether its sheep, cattle, buffalo, chickens, bees or pigs – if animals bring you joy, a farm will give you plenty of room to share with them. Or perhaps you're more of a native animal lover? Certain properties attract wildlife ranging from echidnas to wombats, platypus to kangaroos – you'll just need to plan your fences to protect your veggie patch!

You're a foodie

If you're a regular shopper at farmers' markets and love sourcing fresh produce and unusual ingredients – and want them growing within a few steps of your kitchen door – living in a rural community will give you access to amazing produce. From trading with neighbours to planting your own food forest, there's nothing quite as satisfying as serving up and eating food from your own patch.

For a sense of personal security

In this age of global upheaval, terrorism and mass migration, a desire to 'get out of Dodge' is taking hold for some people. Maybe you just want to be more in control of your surroundings; being away from high-density areas brings a sense of comfort for those feeling vulnerable. Or perhaps you want to go off grid and minimise your exposure to technology, corporations, government and systems.

People who are aware of global and local risks are looking to rural areas to cushion the impacts of a changing world. For example, according to an NRMA report in 2013, Australia would run out of petrol in just three weeks if oil imports were interrupted. Imagine the chaos in the big cities and across the country if for some reason freighters couldn't deliver fuel to supply the 25 000 truck trips needed each week to cart food to the population of Sydney alone.

Love of variety

There are very few businesses that will challenge – and reward – in the mental, physical and spiritual ways that farming can. There is so much variety in farming, from interactions with animals to forecasting business opportunities; from use of hand tools to use of social media; use of grey matter and what really matters! It takes a long time to identify and exploit all the opportunities available on a farm, and the variety just keeps coming. That said, you will need to tap into patience because farming takes time. You're operating in the real world where plants and animals don't grow at the click of a button.

You want to make a difference

If you want to make a difference to the environment and peoples' nutrition, farming is a great place to start. There are not many industries you can work in where you exert a real and important role as steward to the land, its people, and creatures. Farmers make a real difference on this planet. It's a calling that matters, a legacy that lasts.

To gain privacy

Sick of people in your face all the time? Are the neighbours too close? Do you want to be able to take in a view without people at your elbows or in your ear? Do you need a fresh start? Are you on witness protection? Farms can offer privacy and a place to get away from people.

A long-held dream

Perhaps you remember the thrill of riding a pony in childhood, or sitting around a log fire as a teen, or the stunning view from a veranda when you holidayed outback. Perhaps you feel there's something missing from your current life, and that you might find it – and more – on a farm. Dreams can come true. Perhaps it's time to turn yours into reality.

There are many reasons people take up farming. What's yours?

2

WHY NOT FARMING?

Why do people give up farming before they even start...or not long after?

The truth is, the bulk of people who dream of farming, never do it. They get stymied by the status quo, have night sweats about all the things that could go wrong (lions and tigers and bears, oh my)! Or they're put off by the perceived downsides of moving out of the city.

"Bushfires can be in the mind of buyers and this can turn people off leaving the city," says real estate agent Paul Clifford of his many interactions with potential tree changers. "But this can easily be mitigated by choosing the right property, clearing around the house and putting in bushfire protection systems."

Clifford says other reasons people give him as to why they don't take the leap include lack of access to public transport and the distance to schooling for the kids. These two factors don't have to be roadblocks to your dream either, as choosing the right location means both can still be within reach.

Finally, bank lending ratios for land zoned *primary production* can be more stringent than standard domestic housing, which means buyers often need larger deposits, or other security, for a farm loan. Given that the cost of rural properties is so much lower than the sky-high prices of city properties, this too might not end up being so much of a problem.

For those who breeze past these headwinds and do start farming, not all of them will end up staying on the land. Here are some of the main reasons people give up farming. It's important to be aware of them before you start, so you can try to avoid, or at least understand, the ramifications of each.

Unrealistic expectations

If you've watched a few 'living in the country' shows, you've probably seen a sanitized, romanticised version of farming where the professional chef/host/model milks a friendly cow before whipping up a delicious custard tart that is enjoyed by beaming locals who've brought along home-brewed beer and elderflower champagne to wash it all down.

The reality is probably closer to this; you had to spend the two preceding days fixing the fence after the neighbour's bull broke in to pay your cow a nocturnal visit, you had to throw away two whole milkings because of suspicious looking brown floaters, and the hay you bought in to feed your cow cost you a couple of hundred dollars, while her vet bills were even more. This probably makes just one bite of your custard tart more expensive than a dessert for four at a Michelin-starred restaurant. Realising this, you take a big slurp of your neighbour's muddy-tasting beer, and when that doesn't work, you down two glasses of the elderflower concoction, which offers less fizz and alcohol than a standard lemon cordial. As the night progresses, squadrons of mosquitoes take over from the legions of flies, and when you finally go to bed after checking on the safety of all the farm animals, you're woken by the rooster's crow at 4 am. That's when you realise you've run straight from the 9 to 5 into the hairy arms of farming's 24/7 demands.

Financial difficulties

Australia has some of the most expensive land prices in the world, which makes farming a tough financial venture from the start. That's why it's important to treat your farm like any other business: avoid overcapitalising and have your head lead the planning so your heart doesn't get stomped on by muddy financial gumboots down the track. If you feel it was easier making money working for someone else, the drip feed of a regular salary might beckon you back.

Financial FOMO

City house prices have been shooting up faster than weeds after a summer storm. The effect of this is that even people who really dislike the changing nature of their life feel too anxious to sell up in fear they will never be able to get back into the city market. This is FOMO (fear of missing out). A way around this might be to keep an investment property in the city as a hedge. It can be a big decision to leave the city, however, it's also a big decision to keep leading a life you don't enjoy.

Age and physical health

The average age of farmers continues to rise. According to the ABS, in 2010–2011 the median age of farmers was 53 compared with 39 for all employed persons. Farming, being a combination of business and lifestyle, means you may get old doing what you love. However, it also means that what was so easily accomplished in your 30's, 40's and 50's might be that little bit harder in your 80's.

The physical health of you or other family members can have a big bearing on your farming longevity. If a relative requires long periods of treatment in distant hospitals, or a serious disease is diagnosed, it might be necessary to move closer to care.

Relationship breakdown

Farming can cause financial and other stresses on relationships. When relationships break down, assets need to be divided. Separation and divorce can lead to forced downsizing and relocation.

Children/grandchildren move away

Younger generations are so much more mobile these days, so not only might your offspring want to move to the closest city when they turn 18, they might end up taking a job in Singapore or Hong Kong a few years later. Sometimes, with the kids gone and the gradual loss of their youthful energy and strength, people find running the farm becomes harder or less fulfilling (or maybe it becomes a whole lot easier, depending on your kids). Although many families will congregate back at the farm for heart-warming holidays, weddings and to create more amazing memories, not being physically close to their children and grandchildren means that some sell the farm so that they have the freedom to chase their children around the country or world.

Nature

Some people have a genuine aversion to snakes, spiders and skittery critters, and even with all the best intentions, never quite find their inner David Attenborough. Perhaps the paved and covered earth of the city, where hardly anything survives the reflected glare off the glass and concrete, will always feel more like home for some people than a place where assorted other creatures want to live.

Exhaustion and care fatigue

Running a farm can leave even the most energetic and upbeat people mentally and physically exhausted. The care fatigue that can arise from running a business involving living things – and the old maxim that 'there is always something to do on a farm' – can leave you feeling run-down and over it. The loss of animals and crops, having to face pests, diseases and nature head-on, and financial stress can lead to anxiety and depression. That said, studies show that farmers are no more likely to suffer from depression than the average person, and that farmers often show great resilience in the face of adversity. One of the problems that can cause burnout is not being able to get away from your farm so you can appreciate it! Weather events, animals birthing and the demands of Mother Nature mean you need to find someone who can take on some of the responsibilities for you so you can get away and have a true break.

～

FARMER SPOTLIGHT: Beef jerky

This young couple found opportunity in their property's isolation, creating a product that is enjoyed around the country.

Farmers: Douglas and Rachelle Cameron
Farming: Crossbred Charolais, Angus and Brahman cattle
Farm name: Nive Downs
Farm size: 13 760 ha (34 000 acres) - combining 2 properties
Average annual rainfall: 525 mm
Where: Augathella, QLD
Farming since: he was born, but on this property since 2005

Douglas and Rachelle both grew up on large cattle properties, and now run an even bigger one with 1 000 breeders and progeny, so why are they in a book on small farming? Because they're a family operation with an inspirational approach to dealing with setbacks, and we can all learn from that.

"We'd had four years of drought and I'd been contracting off the farm doing heavy machinery work. There was no feed and we were looking at selling cows after sales costs of $50 a head," said Douglas. "It wasn't worth trucking them to agents because of the fees and the feed, so I took some myself and coming back from the saleyards I saw a bag of jerky being sold for $5. I realised ten bags of jerky were getting the same price as one whole cow."

"Those downer years made me sit and think about the problems we were facing," said Douglas. "We're on a property a thousand kilometres from the Eastern Seaboard, it's a huge distance to market and there are constraints on fresh produce. We'd had three seasons where we couldn't even fatten a crowbar and we had no control over prices. I also loved eating beef jerky, and am a bit of a foodie, so introducing a shelf stable product aimed at the consumer market just made sense."

Working 60+ hours per week (Douglas 90% on physical labour and 10% on the business, Rachelle 80% admin and 20% physical) they cover jobs including fencing, stock work such as branding and vaccinations, upgrading infrastructure and maintaining water troughs and regrowth. Having worked out the best way to produce the jerky it's now expertly processed by a local butcher. They sell via wholesale, local retailers, online and at special food fairs and markets.

Working a farm can be relentless, but Douglas sees benefits in seeing his kids living a low stress life fishing and yabbying and being able to run without worrying about traffic. He also loves the freedom of getting out of bed and deciding what he wants to work on that day.

"I get fulfilment from this lifestyle," he said. "I went on a breeding program and got into hybrid vigour, and when you bring those calves in now to brand them, you get a physical high seeing them getting bigger and better framed. On the land you get to see your achievements, there's a physical return and you just can't get that from a regular job."

Food for thought from Douglas and Rachelle

What are the basics required for your farming activity?
Lots of people want to get on the land and have a shot so set your goals, act on them and you will get there. But it's not just about knowing farming, it's about bookkeeping, wages, stock flows, profit and loss.

What are the threats to the industry?
Unstable cattle prices leading to extremely high boom and bust times. The prices we were getting back in 1980's we were still getting in 2016 until this latest spike, but in drought times you get paid per head not per kilo.

Opportunities in the industry?
Organics, paleo and being able to confirm the heritage of the meat. People want to know where their food is coming from, where it was made, its story – the whole paddock to palate is where it's at.

What makes a good farmer?
Persistence, passion and a focus on mid and long term goals. You need to be able to adapt to changing physical and financial environments because the 21st century is going to be full of surprises.

Best piece of advice from another farmer?
Keep it simple and save more than you spend.

Worst piece of advice from another farmer?
There's no bad advice, you can learn something from everything and everyone, even if you don't agree.

What would you farm if you weren't involved in your current activity?
Redclaw (crayfish) because I love eating them.

Douglas and Rachelle's top 5 tips
1. Get a lot of info from a lot of people, take it all in and weigh it up. Then, if it makes sense, go with it.
2. Understand your constraints and move with them in a positive way. That means making the most of what you have and going for small wins, not just big ones.
3. Open your mind and work out ways to get out of sticky situations.
4. Dive into the financials and work out your position before you act.
5. Opportunities will arise so start small and work out where you can add value, then build, improve, sell and work your way to a block you can be proud of.

What's the best way to enjoy your product?
Just eat it. It took me five months of experimenting to get these products right and they're a real journey in your mouth. There's Original, Heated Garlic, Thai Fusion and Hot & Spicy, and I'm working on some more!

Difficult neighbours

They might be a kilometre away and you might not see them when you shut your front door or drive out of your driveway, but difficult neighbours can make your life just as much a misery in the country as in the city. If you can imagine a fencing dispute on a quarter-acre block, imagine it on a 100 acre one! If you get serious about a property, be sure to pop in to meet your prospective neighbours on all sides to get a feel for who you might be sharing your boundaries with. Don't just drive by in the daytime, either. What sights, sounds and smells emanate from the surrounding properties at 5 am? At 7 pm? At 11 pm?

Isolation

A feeling of isolation in the early stage of a move to a new place is normal, but if that feeling continues for years – and gets stronger – it will set up a longing to return to your old home base. It's not just about physical distance from people and places. Perhaps you miss the camaraderie and noise of a busier workplace, or the warm feeling of just being able to fit in with old friends. To get comfortable, you will need to overcome the mental distance you feel in your new community. You can do this by immersing yourself in local groups such as book clubs or the bushfire brigade, by speaking with local business owners, or by volunteering and attending as many community functions as possible, from dances to fundraisers. Taking the time to explore your new area like a tourist will also connect you to the land and its highlights, and planning visits to and from friends and family will also help.

Environmental change

You move to the country for the fresh air, gorgeous views and unpolluted waters...but oh dear, that seems to be where mines like to start up! Mines can move in, groundwater can be polluted by run-off from air bases, or a bushfire could decimate the forests nearby. These are just some of the environmental changes that might make you consider a move.

Council zoning changes

Council can also change the rules. Perhaps you bought a farm zoned for intensive agriculture, but a council zoning change means that suddenly you're smack bang on the border of a large housing development...and all your new neighbours now complain every time you start your tractor, fertilise your paddocks or one of your animals' moos, baas, or crows. Maybe the defence department wants to expand, or the government scopes your back paddock for a new runway. Or perhaps everything is going great and you're loving the life, but a developer trots down your driveway with a big dollar offer to subdivide and develop your land. What do you do?

Lack of planning by tree changers

Dr Angela Ragusa of Charles Sturt University has done social science research into tree change and lifestyle migration and discovered there is often a lack of planning done by people prior to relocating, other than going on a holiday to the area prior. She says renting a place to try-before-you-buy could help you discover if you are socioeconomically/politically/religiously likeminded with those living in the area. This can be tied to a positive experience, as is being in a place where you will have other tree changers to befriend. Dr Ragusa highlights the need to think long, not just short term about your mental, physical, economic and social well-being needs rather than romanticise that all will be better by making a 'lifestyle' change. "It will likely be very different, but new annoyances and problems quickly replace old ones," she said.

The bucket list

Like parachuting, climbing Everest and swimming with whale sharks, maybe owning a farm is just another thing to check off your bucket list. You might farm for a few years or a few decades, but you find you can't keep running the farm AND get to all the other things remaining on your bucket list. In which case either the farm or the bucket list will have to go.

In summary, one or more of these reasons might play a part in people giving up the farming life, but in the end, does it really matter whether you stick with farming or give it up? No! That's because when it comes to taking on a new venture such as farming, it's the journey and exploration that often makes us happy, not the destination. So, don't be scared about giving farming a go, even if you don't end up staying locked to the land. It's following your curiosity that makes it so worthwhile.

3

HAVE YOU GOT WHAT IT TAKES TO BE A FARMER?

I f we can go from clicking a keyboard to clicking the shears, from not knowing the difference between good bugs and bad, or between weeds and wheat, so can you. It's amazing what you can achieve when you have the drive to make it happen. But you need more than drive; you're going to need a combination of eleven important elements to make it as a farmer.

Land

A farmer needs access to land (or water if you're going into aquaculture), but that doesn't necessarily mean you need to acquire it outright. Think laterally; some of the various options to consider include the following.

- Rent multiple suburban backyards in which to grow produce, and in doing so acquire a loyal customer base for your produce.
- Caretake a farm in return for the right to cultivate crops or agist stock on an agreed portion of the farm.
- Work with another grower on their farm, in return for a wage and a bed or two of your own.
- Lease a farm or part of a farm.

- Share the purchase of a farm with family and friends.
- Buy a farm outright.

FARMER SPOTLIGHT: Rhubarb

Not owning farmland hasn't stopped this farmer from creating a profitable, niche business.

Farmer: Malcolm Ryan
Farming: Rhubarb
Farm name: Rhuby Delights
Farm size: 0.10 ha (¼ acre) rented
Average annual rainfall: 946 mm
Where: Don, near Devonport, TAS
Farming since: 2014

Malcolm grew up on a 488 acre dairy farm behind Burnie. His mum always had a row of rhubarb in the ground and would prepare tarts, sponges and even champagne with the perennial plant. As a young boy, he'd forage through the plants for the frogs he needed for fishing, and after years travelling the country working in roles as diverse as grape picking, cotton harvesting, mining, plumbing, milling and as an alderman on Burnie Council (where he fought against managed plantation schemes being allowed onto good food production land), Malcolm had a lightbulb moment.

"I wanted to make rhubarb new again. But I had no land, no facilities and everyone was saying I wouldn't be able to do it. Then I got a break. I was out helping prune a couple's fruit trees and we got to yacking and they offered to let me grow rhubarb on a small patch of their farmland."

In the early days, not owning the farmland or commercial kitchens where he prepared his product freed him up to invest in research and development. It also showed him the benefits of collaboration with other organisations and he contracted a social enterprise to help pack the product. The only major item he purchased was the special chocolate-coating machine, imported from Italy, that turned his organically grown, freeze-dried rhubarb into a high-end treat.

"One of the biggest hurdles was working out how to coat the rhubarb. So many chocolatiers said it couldn't be done, but I thought laterally and compared it to how I used to bale hay. The principles are basically the same: you need the right temperature, the right moisture content, not too much humidity or rain and the right machine and operator for the job. Done!"

Malcolm's typical 60+ hour week (8% physical labour and 92% the business) includes management of the crop (watering, dead leaf removal, weed management and harvesting of up to 100 kg at a time) and production and selling of the product. The product is now stocked in four states and is sold online and with a big focus on special events and large fairs. He is a big believer in farmers' markets, but doesn't attend them himself as he no longer wants to work all week and on weekends. Malcolm has now expanded the business, setting up a manufacturing centre where he now also coats strawberries, blackberries, black currants, coffee beans, sultanas and even makes a hot chocolate and rhubarb product. He has reps helping get the product out across the mainland.

"I'm glad I didn't fall into the trap of thinking I had to go big to supply the chains. I can grow a lot less rhubarb and value add it, and I don't need to worry about paying off expensive land or commercial buildings. I suppose I've found a way to stay in touch with nature, challenge myself and have fun at the same time."

Food for thought from Malcolm

What are the basics required for your farming activity?
Understand the plant and its requirements. Rhubarb is a copious feeder and needs massive amounts of organic matter and mulch. But it also needs a tough environment to get that deep red colour, so you need to balance on that fine line between crop loss and great produce.

What are the threats to the industry?
Many people haven't been exposed to rhubarb and don't know what it is, so a lot of education needs to take place.

Opportunities in the industry?
Rhubarb's a healthy product with plenty of Vitamin C and antioxidants, so it's a great plant to be involved with. There are so many ways to value add it as well, and for farmers in Tasmania, the positive image of produce from our island is a major opportunity.

What makes a good farmer?
Observation, for starters. When I was growing up, most farmers sprayed out the chicory, plantain and dock, but my dad noticed the calves never went for green grass, but would hang at the fence lines, nibbling at herbs and weeds. He'd cut extra for them, chop it and add it to their feed to keep them healthy, and he saved money on chemicals. Also, the philosophy of leaving the farm in a better state than what you got it in; it's about farming and living sustainably.

Best piece of advice from another farmer?
Think outside the square and value add.

Worst piece of advice from another farmer?
Don't do that, it won't work.

What would you farm if you weren't involved in your current activity?
Hemp. It has so many uses, is easy to grow organically and is a great alternative rotation crop.

Malcolm's top 5 tips
1. Collaborate with other farmers and businesses.
2. Innovate...don't just do what's being done, take it as far as you can.
3. Manage your work/life balance so you don't burn relationships.
4. Listen to ideas but tune out the naysayers.
5. Don't try and chase better and better returns like corporate farming – find a balance you're happy to live with and enjoy life.

What's the best way to enjoy your product?
So many ways! A rhubarb tart with shortcrust pastry washed down with rhubarb champagne, followed by some Rhuby Delights.

Money

Like any business, there are going to be set-up costs. Even if you don't need to buy the land outright because you've set up a land-sharing agreement, you will still need start-up funds for tools, inputs, insurance, transport and to replace the income you would have been earning if you were employed elsewhere. In addition, you'll need a safety net of cash in case crops fail. This is why some people start farming in a small way, working at another job while they begin to develop the land.

In addition to your own money, you might be able to tap into start-up funds, loans and grants through government organisations operating in the region. This can be a huge help for a young enterprise, but requires quite a commitment on your behalf, including extensive paperwork and recordkeeping. Further, the money might not come through as quickly as you need it, so having access to cash is crucial before beginning a farming venture. There is also the potential for crowdfunding, which is discussed later in the book.

Good health and physical strength

All farms require physical labour, though it varies depending on what you are farming. Farming edible insects will be less strenuous than farming bananas, and although these days you can choose a ride-on mower over a scythe or a rototiller over a hoe, you're still going to need to be able to lift and carry things and be active on your feet. If you're not physically able, you'll need to be able to earn enough from your venture to employ someone to do all the manual tasks required on a farm.

Time

It can take just a few hours on the stock market to win or lose a fortune, but in farming it takes years. Having time on your side is key to making a go of a farm, and for some families it will be the next generation who reap the benefits of your labour. However, time is not just about how many hours you have left on this planet. It's also about how much time you can devote to the farm (e.g. if you're raising kids, this will eat into how much time you have available), and how much time Mother Nature needs to produce the goods (there's a big difference between the turn-around time of baby spinach leaves and that of giant bunya nuts). Just the nature of farming takes up loads of time (there are animals to feed, sunsets to soak in, fields to wander), so you'll need to be able to pare back other commitments (such a television viewing!) to make a go of it. If you're 67, rather than 27, consider buying a farm that is already estab-lished with mature trees so you don't need to wait until you're 87 for your first harvest...then again, half the fun is watching things you planted grow!

Energy and stamina

If you're a couch potato, you'll never be king of the potatoes. Self-motiva-tion, combined with mental and physical energy, provides the carbohy-drates you need to fuel your farming ambitions. You'll also get extra motivation from seeing the fruits of your labour.

Training and knowledge

Those who have formal agricultural training and/or previous experience are already one step ahead, but if you're totally new to farming, your willingness to seek guidance and implement best practice shows that you too have it in you to be a farmer. A voracious appetite for books, workshops and podcasts, along with courses, mentors and long chin-wags with other farmers are stepping stones in the right direction. Look for local groups like Landcare and TAFE to put you in touch with local farmers and conditions. There is more on this subject in Chapter 10, *The getting of wisdom*.

Usefulness

These days, many of us are useless. We haven't been taught the corner-stones of life like how to grow things, make things and fix things. Super-market chains, cheap goods from China and a society that sees nearly everything as disposable just don't provide the best training grounds for farmers. If you already have a problem-solving nature or a fix it rather than throw it away attitude, that's great. If you don't, you're going to have to create a whole new recycling, upcycling, handyman/woman persona!

Wide-ranging business skills

Even if you have the greenest green thumb ever to sprout on planet Earth, you're still going to need business skills. Marketing, sales, budgeting and recordkeeping are all needed in farming, so you'll need to show a willingness to learn...or to outsource these vital tasks.

Access to help

It can be done, but it's very hard to run a farm on your own through all four seasons. At our place, our children helped from the start (yes, in some countries it's called child labour), and at times, especially during harvest, you might need to call on other family members, local experts, backpackers, neighbours and casual staff.

Personality traits

A University of Queensland study in 2000 by Dr Marilyn Shrapnel, Dr Jim Davie and Dr Bruce Frank of the School of Natural and Rural Systems Management, funded by the Australian Research Council, studied 60 long-time farmers involved in larger-scale grazing and mixed cropping in central Queensland. They determined that these farmers exhibited just five of the 14 personality types psychiatrists recognise in the general population. These personality types were:

- a capacity for hard work and perseverance (conscientious personality style)
- autonomy, capacity to make decisions (vigilant personality style)
- great capacity to cope with adversity (serious personality style)
- comfort with solitude, fully self-contained, stoical (solitary personality style)

- comfort with a small circle of friends, with little need for company (sensitive personality style).

But what about farmers on smaller farms, and those new to farming? One thing's for sure; successful farming demands more than a clock-on, clock-off attitude. If you have a mix of the following personality traits, you'll be off to a great start:

- positive, can-do, will-do attitude
- adaptability
- resilience
- resourcefulness
- attention to detail
- patience
- problem-solving skills
- self-control
- enthusiasm.

Passion

Passion is the energy that leaps white hot from the furnace of your being. When you are passionate about something, you are driven to experience things, do things, achieve things. Passion makes you fierce and unstoppable. It both exhausts and feeds you, draws from you and gives to you. It's the hot spot in your heart that helps you to go on when others would give up. Passion makes us feel alive, and when you are passionate about the land, it's an incredible exchange between the earth and a human being that can lead to a well-lived life.

So...have you got some or all of what it takes to be a successful farmer?

Young farmers

Young farmers face some specific challenges, so we spoke with Joshua Gilbert from Guraabi Downs Brafords. Josh was Australian Geographic Young Conservationist of the Year 2016 and is a Worimi man.

What do you think is the biggest challenge facing young farmers?
Accessing capital to purchase their own farms. In order to get started on your own patch of ground, you have to find something big enough to pay the bills, come up with the deposit and be able to convince the bank to lend the money. Farms are just like any other business, though they can seem a little more romantic, and this means there's not always the same playing field between getting a loan for a coffee shop versus a loan for a farm to run cattle. There are investors and companies starting to bridge this gap, and other options like leasing and share farming are becoming more common.

What mistakes do you see people making again and again?
The greatest mistake I see is people not knowing where to start. They buy a bit of land, yet the endless possibilities of livestock or produce makes it hard to work out the next step. I'd recommend doing some of this homework before buying the block and setting the intentions early on what you hope the farm will achieve. Then, find a mentor. Find someone in the area who can come out and help with cattle work when needed or give advice. A trip to the local produce store will usually help identify someone who can help.

What do you think are some great opportunities on the horizon for small farmers?
The opportunities for small farmers are much greater than the fence boundaries. There is so much potential in farming and there are many incredible farmers you can learn from and work with. The ability to connect in with other cultures, build connections and collaborations, and share is truly at the core of our farming past in Australia. Ensure you think out of the box, and bring your own unique skills, passions and interests to agriculture – the industry needs and welcomes it.

What would you say to a young person thinking about moving to the country?

Do it! Whether you want the opportunity to connect back with land, be able to eat some of your own produce, share your patch with others or retreat from the hustle and bustle, the country provides the ability to do all these things. Before signing up though, think of the practicalities – who will help your cow calve if she is having trouble, who will water the gardens or fix a fence if need be?

Anything else you'd like to weigh in on?

By starting your own little farm, you have a responsibility to look after the land and keep it thriving for future generations. Grab a shovel, dig in and share the passion.

4

THE CHICKEN OR THE EGG...OR THE AREA?

What comes first, choosing *what* to farm, or choosing *where* to farm? Do you choose the animals, plants and produce you are passionate about and from which you can make money, and then find the perfect place to nurture them? Or do you find a region and piece of land that speaks to your heart, and then choose the produce to grow on it? With two such distinct ways into farming, this is the 'chicken or the egg' scenario of your agricultural adventure.

Our family chose the area first, then the farm, and then – after much experimentation – what we would farm. This probably isn't the way to do it if you want to be commercially viable fast! It did allow us to explore our interests and passions (that's the 'making a life' part of making a living), but it also meant we had to work incredibly hard to make it pay.

For people who are already on the land, you might have chosen your farm because it was perfectly set up for your intended venture, it was handed down to you through family, it was what you could afford at the time, or you too might have fallen in love with a region and begun the process of shoehorning the produce to fit it.

If, however, you're lucky enough to be just starting out, choosing what to farm and/or where to farm are critical decisions that will affect the course of your life.

Often, when people are faced with big life decisions they settle for the status quo and choose not to make any decision at all. You might know these people; they're the ones still dragging themselves into work every week where they hate at least 2 000 of the 2 400 minutes they're at their desk, tools or workstation. They're the ones who rarely feel joy, the ones who know there's something more out there – a completely different way to live – but who feel paralysed about moving towards it. Let's hope that's not you! Let's hope you totally get that you only have one life to live, and if you're going to make your farming dream a reality, you're going to have to actively work through how to make these decisions.

So, how do you make such a big decision about the *where* or the *what* of farming? It's made easier by looking at your options based on the following criteria:

- Location: your willingness and ability to relocate.
- Passions: what can't you stop thinking and dreaming about?
- Personality: your personality type and traits.
- Finances: your financial situation and expectations.
- Skills: your education and acquired business, life and/or agricultural skills.

Location

What you can farm is often dependent upon the region to which you are prepared to or want to move. That's because certain types of produce flourish in certain areas due to climate, soil conditions and other environmental factors.

If you can't see yourself moving more than one hour from Melbourne's CBD, it's unlikely you'll be able to become a coffee baron, although you might be able to grow salad microgreens on a small acreage reasonably close to the metropolitan area. If you're in the steamy tropics up Cairns way, you'd be biting off an insurmountable challenge trying to grow Pink Lady apples, but you'd be in a great climate zone for growing giant granadillas.

Even within a given municipality, adjacent farms will have vastly differing abilities to support different types of agriculture. This can be because of their unique topography, land size, access to water, soil type, microclimates, existing infrastructure, ease of access and past use.

Thinking about where you might relocate to isn't just about choosing a location where the climate and soil will support what you want to grow, though; it's made much more complicated by personal, non-farming considerations.

These include:

- your ability to move away from established support networks
- the desire to remain near or move closer to family and friends
- the proximity of the area to existing or potential off-farm work opportunities
- the need to be near schools (if you have or are planning to have children) – or being prepared to consider boarding schools
- how close you want or need to be to medical services
- the expectations placed on you to care for ageing relatives
- a preference to remain near highways, train lines and regional airports (especially if you're an avid theatregoer or shopping mall aficionado, or you simply can't do without a regular dose of the big smoke)
- easy access for visitors.

All these factors play a role in determining where you will locate your farming venture, and therefore what you will be able to farm. It's likely you will end up choosing the location of your farm based on a mix of commercial and personal reasons. This is where your passions come into play.

~

FARMER SPOTLIGHT: Walnuts

Jane, a nurse, and her husband Phil, a rehabilitation counsellor, dreamed of living a rural lifestyle. They moved from Sydney to Tasmania in 1986 and 11 years later planted 1 500 walnut trees, which now produce award winning walnuts.

Farmers: Phil and Jane Dening
Farming: Walnuts
Farm names: Coaldale
Farm size: 12 ha (29.7 acres)
Where: Coal River Valley, Richmond, Tasmania
Average annual rainfall: 520 mm
Farming since: 1997

When the Denings first moved to Hobart, they continued working in their professions, but eventually realised that by adding just 20 minutes to their commute they'd also be able to secure the rural lifestyle they'd always wanted. They read extensively, keeping their ears and eyes open for opportunities. They sought advice from a company developing the walnut industry in Australia, and when the right block of land came up with river flats, irrigation potential and low rising hills, they purchased it. They planted grafted varieties of Franquette and Chandler and now produce fresh in-shell walnuts, vacuum packed kernels and jars of pickled walnuts.

"Even though we've worked off-farm for many years, we've always treated the farm as a business, not a hobby," said Phil. "We've focussed on top quality product and building brand awareness, and we developed our independence by building and owning our own processing equipment."

The Denings are actively involved in their community, and are always open to trialling different methods on their farm.

"For many years, we've used an orchard floor management system where we mow the inter-row grass and throw it sideways under the trees for mulch and to suppress the weeds. We had been thinking of reducing our carbon footprint by reducing the mowing and using sheep to keep the grass and weeds down," he said, "but to do that we needed to change from a floor-based sprinkler system – to avoid the sheep kicking sprinklers around and pulling out tubes – and convert to suspended lines and sprinklers."

Then the drought hit and the Denings brought their trial forward so they could help a neighbour by taking 150 of his sheep onto their land. So far, they're happy with the trial, and the bonus of masses of sheep poo helping to naturally fertilise the orchard.

Phil typically works about 30 hours in the business a week, while Jane works about 10 hours, as she still works off-farm as well. Their week involves orchard work such as irrigation, mowing, fencing, pruning, spraying and repairing equipment, as well as managing stock, staff, online sales and packaging.

"We might have been better off financially had we stayed living in the city, and our children might not have had to move away for their careers," said Phil, "but we get so much satisfaction from being good at farming and from being a part of a good community. We absolutely love the outdoor lifestyle, the autonomy and the variety of activities that go into farming. We're living the country lifestyle we dreamed of."

Food for thought from Phil and Jane

What are the basics required for your farming activity?
Land, water and a long time frame. You also need to remain open to new ideas, whether it be about land management, processing, packaging or marketing.

What are the threats to the industry?
There aren't many threats, as the industry is in a growth phase domestically and internationally.

Opportunities in the industry?
Walnuts are a stable and robust product, which makes them generally easy to manage, store and transport. Demand is high and walnuts are a desired superfood, so opportunities exist for high-end sales of top-quality products.

What makes a good farmer?
Someone who's practical, pragmatic, resilient and resourceful. Passion is crucial, but you also need sufficient capital to get the plan going, good problem-solving skills and a long-term view.

Best piece of advice from another farmer?
Be as self-sufficient as possible with knowledge and equipment.

Worst piece of advice from another farmer?
We've never really received bad advice.

What would you farm if you weren't involved in your current activity?
Another niche horticultural product we could add value to such as juniper berries.

Phil and Jane's top 5 tips
1. Research the market well
2. Write a comprehensive business plan
3. Have a strong marriage
4. Keep learning
5. Engage in your community and join groups

What's the best way to enjoy your product?
Walnuts are a true health food and can be used in countless ways from a walnut, beetroot and feta salad to a vegetarian meat loaf.

~

Passions

What are you passionate about? It's important to identify your passions to help you shape the *what* and *where* of your farming future.

For example, if you are truly passionate about beer, a big move to Tasmania to start a hops farm (with onsite tasting!) might meet your personal and commercial goals.

If you're addicted to arts and crafts, a lavender farm where you can produce all sorts of value added goods such as potpourri, flower arrangements and lavender-stuffed bears might fulfil your 'making-things' side.

If you truly love cheese but don't want to move far from home, an artisan dairy on the outskirts of a major centre might offer the perfect platter of opportunity.

Just as your passions can enhance your farming experience, they can also make it unbearable. Let's say you really, really love animals. It could become a terrible burden for you when the time comes to home-kill them or send them to the abattoir. Even having to put sick animals down can become too much for some people. If this sounds like you, you'll either need a vet on speed dial (and the budget to pay for it), or you might be better off farming plants. And let's say you currently don't go a day without surfing/shopping/sipping tea with your sister...a move to far-western Queensland, a long way from all your big loves, might just end up being too much to bear.

So...what do you really care about? How might your passions fuel or complicate your choice of location and produce?

For example, if your kids are your passion, you might want to give them an amazing childhood, but also be within 30 minutes of a school and doctor.

If you love overseas travel, you might not want to be tied down all year round, so you could choose seasonal produce with export opportunities to enable you to still enjoy your annual globetrots.

Or perhaps it's extended and/or ageing family members who mean the world to you, so you'll want to look at areas serviced by country trains that run at least twice a week. You might also look at a property with separate accommodation or extra bedrooms.

Jot your thoughts down now to see how your passions might impact (both positively and negatively) on the *what* and *where* of your farming dreams.

Personality

In addition to your passions, think about how you can align your *what* and *where* of farming with your personality. Basically, you want a good fit between you, your farming venture and its location, so that your days are a joy.

- Are you risk averse? If so, it might be best to choose land near your current location, along with produce that can be farmed part-time, so you can keep your current career until your confidence and farm income take off.
- Do you love the outdoors or are you more comfortable inside? If you're happier under shelter, perhaps you should investigate types of farming that can occur undercover such as growing seedlings in nurseries, hydroponic operations in

glasshouses, or mushroom cultivation in sheds. These types of farming also offer lots of flexibility in terms of location, so this might be a great way to have your produce and eat it too!

- Do you like being around people? If so, you might want to steer clear of the back of beyond and explore more populated rural areas with vibrant farmers' markets and active communities.
- Are you comfortable being on your own? This means you might be able to find a farm that is out of the way, a peaceful haven that is quiet and well-priced.
- Are you a worrier? Maybe you can try out different types of farming as a volunteer, or take a few farmstays before taking the leap. Talk to the owners about their experiences and speak to farmers at markets.
- Do you hate feeling tied down? Perhaps you should avoid farming animals that need daily care, and instead consider farming perennial plants. They need daily care too, but won't tax you as much emotionally as the need to always be there for animals. It's also easier to find someone to take over the care of the farm when you need a break.

Sometimes, what you think you might want to farm could clash with your personality type. Let's say you're passionate about fine food. That doesn't necessarily mean your personality type will be gratified by building compost piles of steaming fresh manure in which to grow it. Still, maybe the pride you get from creating a nutrient-rich and biologically friendly product will help you get over any obsessive-compulsive dirt-avoidance tendency.

Finances

We've talked about relocation, passions and personality, but your financial situation will likely be the key factor determining where and what you will farm.

In relation to the *where*, different regions come with widely varying price tags. To give you an idea of the range, just click on a real estate site and compare the different prices per acre for farmland in Byron Bay, Blayney, Barkly, Bajool, Burnie, Buladelah, the Barossa and Balingup.

Certain areas also offer better off-farm income opportunities that can supplement your lifestyle while you set up your farming operation. Within each region there will be all sorts of real estate choices to make, all at different price levels. One thing people new to farming often don't realise is that although buying vacant farmland is so much cheaper than buying land with infrastructure in place, it can often end up being much more expensive. Yes, you get to start with a blank canvas, but it might end up costing you a Van Gogh!

In reality, something that looks like a bargain at $400 000 might end up costing you $800 000 by the time you add power, a gravel driveway, sheds, a house, dams, irrigation, tanks, yards, trees, gardens and fences. It will probably also take a few years to get approvals and everything set up and built before you can even start your venture, so this could add another $100 000 or more in opportunity cost. Buying a property for $600 000 that's already been developed might suddenly seem like a bargain!

Not everyone has that much money to invest, so the good news is that buying is not your only option. Perhaps you could lease a farm, or part of a farm, to get your start (this is discussed further in Chapter 5). You might buy an undeveloped block and be prepared to live in a caravan or shed for years while you build your dream. However, if you're going to be running a business, you will still need some infrastructure – and you can't ignore the potential cost of that.

The good news is that not all farming requires a huge investment in land. Beekeepers can have a small base for their operations, as they use trucks to transport hives to other farms and forests. Intensive green-house operations can be run on relatively small parcels of land, and you can be a cattle farmer by agisting your herd on other farmers' properties.

The next thing to think about is the start-up cost of your operation. There's a huge difference between the start-up costs of a thoroughbred horse stud and those of a culinary snail farm. And if you're going to be farming flowers, you can get a harvest in your first season, but if you're growing pecan nuts, it might take 10 or more years before you get your first major harvest. That's just one more reason why your financial situation going into farming will determine the *what* and *where* of your farming adventure.

FARMER SPOTLIGHT: Hereford beef cattle

From a start with a second-hand red esky to an eight-tonne refrigerated Pantech truck, these farmers have built demand for their grass-fed beef through farmers' markets and online sales.

Farmers: Greg and Lauren Newell
Farming: Hereford beef cattle and pigs
Farm name: Linga Longa
Farm size: 200 ha
Where: Wingham, Manning Valley, NSW
Farming since: 2002 on this land

Greg's family had been farming Hereford cattle all their lives in the Dorrigo Mountains, but when Greg left school at 15 to work on the farm, his dad sent him away saying, "What are you doing here? You have an interview in the morning in Queensland." So Greg went and did his mechanic's apprenticeship and then returned to the farm to help out. His dad promptly said, "You're too young. Go and get a job."

"So I worked for 20 years in law enforcement and never went back to the farm in case they sent me somewhere else," said Greg, laughing. "Then I bought my own places, and I've since done the same thing with my own kids and pushed them off the farm so they can go and get careers. That way, they can come back to farming with some life skills."

These life skills have seen Greg and his wife Lauren, a marketer, combine to run Linga Longa Farm. With 120 cows, each with a calf at foot, for most of the year, and 15 sows kept on a separate farm, they've come a long way since the A4 pieces of marketing material they tucked under car windscreens at the local showground when they were first trying to get customers.

"We started by building demand, and then built the operation around it," said Greg. "We did frozen meat, and in the early days there was a lot of wastage of the animal, as local buyers only wanted the scotch fillets and the steaks, and being in a rural area a lot of people had access to friends with beef or could get a quarter of an animal from an acquaintance."

Greg and Lauren looked further afield to Sydney and decided on a nose-to-tail and value adding strategy. Their products now include regular cuts of meat, as well as bone broth, marrow bones, dog bones, smoked goods and sausages. They also offer hams, bacon and pork. They travel to Sydney each week to attend farmers' markets on two consecutive days, which makes the eight-hour roundtrip viable.

Greg likens a farm to a demanding woman who needs looking after. "She can be cranky, but you need to be at one with her. You need to say 'yes' to her at certain times and 'no' at others, such as deciding if the pasture needs water or if it should be saved for the stock. You need to prioritise, as you're both master and servant at the same time."

Working 60+ hours per week (Greg 90% on physical labour and 10% on the business and Lauren the reverse), involves fencing, irrigation, pasture and livestock management, attending farmers' markets, packaging, marketing and organising online orders. They have also renovated a cottage on the property for farmstays, which provides credibility for the meat business and offers interaction with customers and another income stream.

"I'm proud to produce food. When people give me feedback and tell me what they've cooked or how much they enjoyed it, it's as though I've sat at the table and had dinner with them. But what I love most is farming with my wife and family. That's what I love, spending time together."

Food for thought from Greg & Lauren

What are the basics required for your farming activity?
The Meat and Livestock Authority (MLA) provides plenty of training, but selecting your farm is the most crucial thing. Look carefully at the infrastructure. If the stockyards are run down, there's a minimum of 20 grand you'll have to spend just for that. Then add in six grand for a crush, not a flash one, just a get-by one. Then there's the concrete pad, the $10 000 for 1 km of fencing, plus, plus, plus. Nothing's cheap in farm-

ing, so put a few more grand in your bank account, because there's never enough.

What are the threats to the industry?
Fluctuating prices, massive amounts of compliance and paperwork. Unfortunately, farmers are the only people who pay retail prices for everything and sell at wholesale prices.

Opportunities in the industry?
Direct marketing of quality products to customers.

What makes a good farmer?
Someone who's calm and thinks twice, someone who has empathy for animals and people, and someone who has good communication skills for building customer relationships. It's also someone who is a part of and cares for the community, such as volunteering in the local fire brigade or ensuring you don't stuff up the creek for other users downstream.

Best piece of advice from another farmer?
My grandfather once said, "You may not love what you do all the time, but what you love doing is what you're doing all the time." Remember that when times get tough.

Worst piece of advice from another farmer?
It's easy money.

What would you farm if you weren't involved in your current activity?
Pre-bottled beer.

Greg and Lauren's top 5 tips
1. Morals, ethics and credibility are everything, so be loyal to your breed.
2. Get into farming as early as you can, choose it as a career and keep up with advancements.
3. Go into farming with both eyes open. Don't believe the glossy real-estate brochures with their inflated stock-carrying capacities. Do your research

through the Department of Primary Industry and the MLA and local land board sites. Check facts through documents and with the neighbours (but keep in mind the neighbours might have an eye on the place too!). Be aware that ads will say everything is closer to town than it really is, and that all 'services to the door' might be a few kilometres up the road. Don't pick a block just because it looks pretty today; how will it be in drought? There might be an embargo on the river, so think about how you'd cope.

4. Restaurants want to pay you nothing for product, so research how you will sell it first, and be prepared for the downsides of farming, the emotional difficulty and ethics of having to cull animals, and the 3 Fs: flood, fires and finances.

5. Have a sense of humour. If you're dealing with cattle, remember they can be like some women, cranky once a month, with a nine month gestation, and can kick and bite without notice.

What's the best way to enjoy your product?
With family around the table.

~

Your skills

What are you good at? What kind of agricultural venture will be best served by your years of education, business and life experience?

Before moving to the farm, I had zero agricultural skills; I couldn't even keep pot plants alive. But my previous experience in marketing and the internet, combined with a driving commitment to learn, meant that my lack of agricultural knowledge didn't become a roadblock. Our venture also succeeded on the back of Andrew's general DIY skills, his hospitality and general business management experience and his never-used degree in animal husbandry. It's probably the diversity of our pre-farm experience that enabled us to envision and end up running such a diverse operation.

What experience, education and skills will you be bringing with you to your venture? How might they save or make you money? How might they determine *what* and *where* you farm?

Handyman skills come in handy on any farm because there's always something to fix, build or tweak. If you don't have these skills, a less isolated property means you will be closer to the tradespeople you will need to help you out.

Likewise with mechanical skills. If you're planning an operation that will make use of tractors, trucks, or milking machines, you'll either need to learn how to keep your equipment in tiptop condition or be close to a town where they can be serviced.

Critical thinking skills will help you solve problems that come your way. Sales skills help in selling and getting the best price for your produce. Human resources skills will benefit you in managing yourself, your partner and your workforce. A background in nursing might be beneficial if you will be caring for animals, and an accomplished home cook or stylist will be able to showcase your produce.

Take some time now to think about your skills and talents and how you can use them to identify what and where to farm. Write them down so you can see how much you have to offer!

In summary, you can choose what to farm first, and then find the region it will do best in. Or you can find a region you love and then farm to suit. To decide what's best for you, consider your ability to relocate, your passions, personality, finances and skills.

5

WHAT FARM SET UP IS BEST FOR YOU?

O ld MacDonald had a farm...but the song gives us no clues as to what kind. Did he own it or lease it? Was it a hobby farm or a commercial farm? Was it intensive or free range, urban or outback, or was he living in an intentional community? Was he a forager?

There is an incredibly diverse array of farming options available in Australia that can provide an entry point for most people, even if you don't have a tractor load of cash.

Lifestyle/hobby farming

This is the type of farming where you support the farm rather than it supporting you. It's the type of farming some retirees pursue, as well as people with full-time jobs who want to retreat to a farm for their weekends. This type of farming does not qualify you as a primary producer according to the Australian Tax Office (ATO), but it does qualify you to keep fit, relax and soak up some privacy and fresh country air.

Primary production

What follows is the ATO's definition of primary production. It comes from subsection 995-1(1) of the Income Tax Assessment Act 1997, which defines 'primary production business' as carrying on a business of:

(a) cultivating or propagating plants, fungi or their products or parts (including seeds, spores, bulbs and similar things), in any physical environment; or

(b) maintaining animals for the purpose of selling them or their bodily produce (including natural increase); or

(c) manufacturing dairy produce from raw material that you produced; or

(d) conducting operations relating directly to taking or catching fish, turtles, dugong, bêche-de-mer, crustaceans or aquatic molluscs; or

(e) conducting operations relating directly to taking or culturing pearls or pearl shell; or

(f) planting or tending trees in a plantation or forest that are intended to be felled; or

(g) felling trees in a plantation or forest; or

(h) transporting trees, or parts of trees, that you felled in a plantation or forest to the place:

(i) where they are first to be milled or processed; or

(ii) from which they are to be transported to the place where they are first to be milled or processed.

Are you carrying on a business or is it a hobby?

The ATO determines whether you are carrying on a business or indulging in a hobby through a set of indicators. The main indicators are:

- a business plan exists
- there are commercial sales of the product
- The taxpayer has knowledge or skill.

Other indicators suggesting a business is being carried on are:

- there is significant commercial activity
- there is purpose and intention of the taxpayer engaging in the activity
- there is an intention to make a profit from the activity
- the activity is or will be profitable
- the activity is organised and carried on in a business-like manner and systematically – records are kept
- there is repetition and regularity of activity
- the activity is carried on in a similar manner to regular trade
- the size and scale of the activity
- it is not recreation, a hobby or sporting activity.

Primary producers might earn their entire living from farming, or they might work full-time elsewhere but still operate their farm as a business. The benefits of running a farm business can include being able to claim against income for expenses incurred, a rebate on diesel fuel, and access to certain programs such as being able to average your income and the tax you are liable to pay over five years to balance out the good and bad years. Seek professional advice when deciding if you intend to claim to be a primary producer with the ATO, and understand that rebates and benefits can change depending on the government of the day.

Leasing vs buying

In the early days of Australian settlement, farmland was literally given away. These days, unfortunately, the price of farmland seems disproportionally expensive compared to what you can earn back from farming it. The positive news is that if you already own a property in the city, it's likely you could sell it, buy a farm in a regional area and have some money left to live on while you establish your farming business. But if you aren't already on the property ladder and don't have funds behind you, another way into farming is to lease part, or all, of a farm. Or if you want to be a city farmer, you could try urban farming.

Urban farming

A commercial approach to community gardening in the city is urban farming, whereby you negotiate with home and business owners, and councils, for access to their backyards, empty blocks and even rooftops.

Urban farming enables you to stay in your current lodgings while building a business with an enthusiastic local market. The production of herbs, salad leaves, flowers and vegetables can flourish in backyard gardens, and urban beekeeping operations have been set up right around the country, supplying honey from suburban hives.

If you are thinking of embarking on this kind of business, keep in mind that the plots will need to be close enough to each other to make travel time between them efficient. You might also need to talk to the landowners regarding additional infrastructure, like water tanks and security measures, and an agreement regarding access. Payment might be made in the way of produce or an official lease.

Indoor farming

Are you a control freak who wants to control everything, including the weather? Do you love gadgets and systems and engineering challenges? Do you want to be out of the sun and rain with a roof over your head? Then indoor farming, ranging from greenhouses to aquaponics, might be for you. There are even converted shipping container farms being trialled in some cities!

Many intensive chicken and pig operations also take place indoors.

Migratory farming or foraging

Perhaps you don't want to spend all your time in one place, or don't have the funds to invest in the farm of your dreams just yet. This is where migratory farming might prove suitable. Migratory beekeepers, certain types of fishermen, feral goat musterers and foragers can travel far and wide, returning to a small but well set-up base for handling or processing.

Agistment

This is where you either pay money to a farmer to agist your animals on their land or receive money from other farmers to agist their stock on your land. If you are taking animals onto your farm or agisting your animals on someone else's farm, it pays to have a written agreement in case disputes arise. The agreement would need to cover the expectations of both parties as to the care of the animals, what would happen in case damage was caused by or to the animals while on agistment, how long the agistment agreement is for and what to do in case of emergencies.

Intensive, broadacre, or free range?

How much land do you need to be profitable? How many bunches of flowers can you grow on a hectare, or how many goats could you raise on that same plot of land? How many chickens can you care for free range versus in an intensive barn operation? This is where philosophy, profit and practicality come together.

Land or water?

Not all farming needs dirt. Hydroponic tomatoes, lettuces, micro-herbs and strawberries are produced out of the ground, and fish can be grown in ponds, tanks and dams, or fished with commercial licenses in rivers, seas and oceans.

Sharefarming

Sharefarming is a system whereby a farmer uses the equipment and assets of a landowner to farm, and in return receives a percentage of the profit from the overall farm's operations. It was common centuries ago and in impoverished countries, where it was and still can be an exploitative model. In a modern sense – conducted between informed parties – it can be of benefit to both the farmer owner, who doesn't want responsibility for everything or is working towards an exit strategy, and the share-farmer, who doesn't yet have the means to buy a farm or is seeking not just employment, but also the benefits of wealth creation, increasing skills and better job satisfaction. Sharefarming agreements in Australia are more common in larger farming operations including those related to grain and cotton growing and dairy farming, where the initial capital requirements are extensive.

Alternatively, if you do have the money to buy a farm, you might not have the time or expertise to run it yet, so you could enter into an agreement with a sharefarmer until you are ready.

Another take on sharefarming is the idea of sharing assets such as tractors and harvesting equipment, even dairy infrastructure. In the case of a dairy business, sharemilking is an option for farmers who own their own cows but for whom it doesn't make sense to own an entire dairy. In this case, they can rent the dairy during the time of the day when it is not being used by the owner. This leads to better utilisation of the equipment and rent for the owner farmer, and doesn't saddle the sharefarmer with huge levels of debt.

Intentional/Cooperative/Collaborative farming

This is where sharefarming gets taken to the next level. Imagine multiple farmers sharing assets for life and business. Imagine the pooling of resources, skills, produce and the cooperative marketing of products.

Large co-ops such as in the dairy and cotton industries are more about wholesaling, but what I'm talking about here is a concept whereby a variety of farmers set up on a large piece of land, or adjoining areas. The land is cooperatively owned and set up with central facilities for processing and selling to the public. Each farmer specialises in a certain field (literally and figuratively), creating raw materials that could be value added together, making the best use of the processing facilities, e.g., the cheese making factory could be used by the cow farmer one day, the goat farmer the next, the sheep farmer next, and so on. The milking facilities would be well utilised, with different farmers using it in shifts throughout the day. The beekeeper would supply honey to sweeten the honey yoghurt and use another joint production facility to make balms using beeswax and the oil supplied by the macadamia farmer, who in turn would use the honey to coat and roast nuts. This type of arrangement is difficult to set up and keep going, but has incredible potential benefits for all involved.

A collaborative farming model such as this is being championed by Stuart Chignell in Victoria. Called Hedgerow Farm, the concept is for farmers to come together to lease a large farm at affordable rates over a long term that can be sectioned for various uses.

"Some people just want to do their own thing, and the model allows that, but others will get many benefits from working alongside one another and, I hope over time, working together with each other. Even the most individualistic of farmers can see the benefits of jointly operating a farm shop, for example," said Stuart.

"Running any small business can be a very lonely profession. Collaborative models increase innovation and productivity. Iron sharpens iron, so to speak," he said.

According to Stuart, the barriers to getting a collaboration like this off the ground seem to lie not in the financials of the deal, but in people's emotions and their willingness to give it a try. Do you have the personality to work cooperatively?

So, that gives you an outline of your options. What type of farm ownership and style do you think you'd like to pursue? And what stepping stones can you take to get there?

6

CHOOSING WHERE TO FARM

L ocation, location, location. How do you decide which patch of dirt to target when Australia offers a whopping 7.692 million square kilometres of land to choose from? Sure, you could throw a dart at a map, go for a very (very!) long drive and grid-search the nation to find an area that resonates with you, or spend a gazillion hours on real-estate sites – or you could work your way through the following key factors to help narrow down your *where to farm* question.

Climate

A region's climate is affected by its terrain (flat, hilly, mountainous), altitude (height above sea level), latitude (how far from the equator) and the effects of nearby water bodies (lakes, oceans, seas and wetlands). These factors show through in long term climate studies that track the range and variation in temperature, wind, humidity, precipitation (rain/hail/snow/drizzle) and atmospheric pressure. Gathering raw climate data is a great start, but you also need to be able to interpret the data. For example, lowish rainfall might be fine if you have access to permanent creeks and irrigation rights (and want to avoid humidity problems that can affect certain types of plants and animals), but too low and your venture might literally never make it out of the ground. Alternatively, high rainfall statistics can also be deceptive in that the rain may arrive in downpours across two months and your land might be arid for the rest of the year.

If you hate the cold and think Queensland is the place to be, you might be surprised to hear that the Darling Downs and Granite Region of the Sunshine state often experience freezing temperatures and even snow. And if you want to head south to Tasmania to escape the heat, the central plateau of the Highlands might be one of the coldest areas in Australia, but heatwaves in the valleys to the west of Hobart mean that farms can sizzle through periods with temperatures above 40C. That's why it pays to study climate statistics that reveal long term weather patterns in specific areas. This information will help you make better decisions regarding the potential suitability of a region for your proposed farming venture.

Let's say you want to farm the world's favourite tropical orchid – the oh so aromatic and delicious vanilla bean – you'll either need a well-equipped hothouse or, ideally, secure land in a place like the Daintree Rainforest in Far North Queensland where it's naturally hot and humid. Meanwhile, the spicy delight wasabi is a semi-aquatic, slow-growing brassica in need of a humid but cool climate, so you can see that you'd need to be at opposite ends of the country to grow these crops efficiently and economically. Different breeds of cattle prefer different conditions too. Climate can affect everything from wool quality to the worm burden you will need to manage in your livestock.

Though climate statistics give you a great barometer (pun intended) of weather conditions in a certain area, this doesn't mean that the particular piece of land you are looking at will totally comply. Large hills can cause rain-shadow effects whereby one side of the mountain (windward) will be wetter, while the other side of the mountain (leeward) will be noticeably drier. Aspect is also an important factor, for example, with latitudinal hills, the northern side will be drier and have less vegetation than the damper southern side. This can be important when deciding which crops to grow where. You'll also need to factor in frosts.

Further, on each individual farm there will be distinct micro-climates created by the particular arrangement of trees, terrain and water in the landscape. On our own farm, a 90 year old beekeeper who had worked the area since his youth told us our farm was in a dry spot, even though it was in a high rainfall coastal area. At the time, we'd had a few years of slush due to the 1 400 mm we'd been receiving annually, but years later when dryer conditions took hold, his words rang true. We would see storms split before our eyes and carry good rains to the north and south of us while teasing us with a spit in the eye. We worked to solve the problem by putting in an additional three megalitre dam and a solar-powered irrigation network to troughs and plantings around the farm, as well as planting key trees on contours using a pit and mound technique, similar in concept to the water and nutrient-retention idea behind swales.

Soil

Soil, the life-giving source beneath our feet, is a mix of rock particles, clay and organic matter. It's what plants thrive in and die in. It's what provides the vitamins and minerals animals need to be healthy. It's what you need to be on the lookout for when you decide where to farm. That's right, rather than looking out at the vista, your focus needs to be down below. Is the soil you plan to base your farming on easily eroded or does it provide good water retention capability? Is the soil highly acidic (great for a ripper crop of blueberries but not much else) or alkaline? Is it high in salinity or high in fertility, rich in magnesium and iron or potentially toxic boron? Does it drain well?

There are entire books on soil, and the CSIRO has an online atlas of Australian soils. But for a quick study, there are ten common soils found in Australia, the best ones for agricultural potential being Vertosol, Ferrosol, Podosol and Dermosol. The ones you want to steer well clear of, as they have the least potential for agriculture, are Tenesol and Kurosol, while you might have low to moderate success with Kandosol, Chromosol, Calcarosol and Sodosol.

You can amend soil over time by applying lime and gypsum and adding organic matter. You can add organic matter by mulching, growing leguminous crops and rotating animals through fields according to a managed schedule. You can also improve soil by minimising erosion, creating compost and loosening compacted soil using implements such as yeoman's ploughs. If you start with a good soil, and care for it well, it makes farming so much easier.

Water

Average annual rainfall matters, but not perhaps as much as you think. An example of this is a farm receiving 1 200 mm of rain a year, but with soils that it just sheets off compared with another farm receiving 600 mm a year that grows amazing crops in the dry season due to the retained moisture in better soil. You can also have too much rain if it falls on the wrong type of land, leading to waterlogging, flooding and lost production, or not enough rain for three months and then it all falls in one weekend.

Farms can also access and retain water through irrigation rights, dams, rivers, natural springs, bores and rain collection tanks. In some areas, if you are on or near a main road you might even be able to access town water. Find out if the region you are looking at has a water carrier service (trucks that refill your tanks) if your tanks run dry...although those long, hot showers will come at a cost, as will accidentally leaving a hose running in the garden.

Water is becoming an increasingly complicated area on farms due to legislation that limits the size of dams you can have based on the size of your property, and due to the sale of water rights to the highest bidder. When looking at farms, analyse whether they will be able to support your farming operations into the future or what you will need to do to best manage too much or too little rainfall. These options could include adding drains or swales and implementing erosion-reduction techniques across the farm.

Fraser Bayley from Old Mill Road BioFarm in Turlingah, NSW summed up the importance of water and having adequate backup capital in an honest, "it's not all champagne and sunsets" Instagram post. Even though he and his wife Kirsti run a truly remarkable operation, it's a salient reminder about the importance of water in any farming operation. Thanks for sharing Fraser!

Retreat and regroup. By Friday 10th November my plan says that I'll have 60 (55m) beds in production and an extra 30 beds being prepared for late Summer planting. After failure to establish crops, I've whittled my current growing space back to 29 beds. I'll harvest a few of those soon and keep 20-25 beds going on the water supply available...this is a huge impact on our Summer income and looks like break even with no wage for me. Was I a new grower with all my capital in this one basket, this season would be crippling. I went into this season with inadequate investment on water infrastructure and put most of my chips on good Spring rain that didn't show. I've had my arse shown to me. I'll be sowing a Summer cover crop into the beds that were meant to be veg, that will hopefully get watered in by the 5-10 mm forecast for tomorrow.

Topography

If you are planning on growing wheat in a big way, you'll be needing flat-bread, land not high-top. But if you are into grapevines, your farm will benefit from a nice north facing slope.

A hilly to steep topography will mean you might need specialised farm equipment to tackle projects safely, and this kind of land is more suitable to forestry and livestock than cropping, unless it is terraced.

Proximity

When thinking about the perfect location, you also need to think about these factors.

- Proximity to customers and markets for your produce – how long will it take and how will you cost-effectively get your produce to consumers or wholesalers?
- Suitability for vehicle access via roads and bridges (older bridges might mean restrictions on the size of vehicle that can travel across them). Does it matter if the roads can't be travelled during periods of wet weather or do you need no creek crossings and a sealed road to your door?
- Family restrictions: what kind of schooling, medical access and access to friends and family do you need to retain? For example, does the school bus drive by your front gate or do you need to drive your child 5 km or 50 km every day to catch it?
- Who are your near neighbours? Too near urban development might mean future complaints (about anything from your crowing rooster to the trucks coming to pick up your

produce). However, being in the middle of a major agricultural area might mean spray drift of chemicals from the farms next door, which could ruin your hoped-for organic status.

- Proximity to your passions and hobbies.

Threats in the region

Weeds, diseases, pests, fire and flood are things all farmers need to deal with, but knowing the specific threats in the area you are looking at will help you either avoid making the wrong land purchase decision or give you the knowledge you need to invest in a harm-minimisation strategy. Let's look at some examples.

Parasites

There's a reason why not many large goat and sheep operations are run in coastal areas of NSW, Queensland and WA, where summer rains are prevalent. That's because these areas provide ideal conditions for killer roundworm species such as Barbers Pole (*Haemonchus contortus*) to thrive. If you were planning to run an organic artisan goat or sheep dairy, it would be best to either avoid these areas or factor in the many extra expenses you will incur in managing your flock's susceptibility. These expenses might involve managing worms through selective breeding programs (where only the least susceptible to the worm survive and are bred from). Extra time would need to be spent managing strict paddock rotation techniques. You might invest in off-ground feed (either hand-fed or bush plantings) and natural treatments (garlic/sulphur, apple cider vinegar and supplements). Even then, implementing all these measures may not save your stock. If you're using chemical drenches, you will need to undertake regular worm counts to decide on the best time to drench to avoid drench resistance. Some of these drenches mean that the milk from these animals cannot be consumed by humans for a certain period.

Wild dogs

These can attack livestock. Their numbers can be high in state and national forest areas, and consequently the farms and regions that border them. If you are planning to run animals in these areas, you will need to consider the extra expenses (financial and emotional) you will incur in either stock losses or the dog-prevention techniques you will need to implement. Strategies could include dog-proof fencing, safe paddocks near the house for overnight protection, running guard donkeys or maremma dogs, with your herd, or trapping and baiting (which might have negative consequences for native wildlife). To find out if wild dogs might be a problem in your area, speak to the local government department responsible for wild dog control.

Diseases

Serious diseases of animals and plants can afflict certain regions and individual farms. Speak with the local office of the agriculture department in your state to determine whether a region is suitable for your intended crop or livestock. They might not be able to give you specifics for the farm you are interested in, but if you are planning on farming bananas, is it near an area that has had an outbreak of the fungus Panama tropical race 4? If you are hoping to breed cattle, sheep, goats, deer or camelids (alpacas, llamas, camels) is it in an area afflicted by Johne's Disease? How will you manage these threats (if it's possible to manage them), or is it best to avoid them by purchasing elsewhere? For example, let's say your dream is to rear miniature horses, but the farm is close to a flying fox colony known to carry the killer Hendra virus. The bats also come to the farm to feed on nectar from the eucalypts and the old mulberry trees dotting the property. You can manage these risks by vaccinating the horses, ensuring troughs aren't left under trees where bat droppings might fall, fencing off trees and stabling the horses from dusk until dawn...or you could set your farm up in a region where Hendra isn't a threat.

Land use

It might surprise you to know that not all farms are approved for all types of farming. An intensive chicken farm doesn't need a huge land area, but council and your neighbours might block your venture if the land isn't zoned for commercial farming. They can also block it if the land is too close to a waterway that could be threatened by effluent. There will also be neighbour concerns about noise, smell and vehicle movements. Find out what your potential farm is zoned for by speaking with the local council.

If you are interested in agritourism, is your farm in a zone that allows for tourism? Think also about the nitty gritty details, for example, if the farm is located on a dirt road, tourists in hire cars would be voiding their agreement with the rental company if they were to drive down it. If you are on a busy road, will you be allowed to have tourists enter and exit your property, or will it be deemed a traffic hazard?

Communication and technology

If you're not planning a hermit's life, before you put down the deposit for a farm investigate internet connection and speed, mobile phone reception and landline, postal and parcel delivery services.

Area crime

What are the crime statistics for the area, and what kinds of crimes are taking place? Are the statistics tracking up or down from past years? As in the city, certain country areas are hotspots for crime, while in other areas you feel safe enough to leave every door unlocked – even when you're away! Speak with the local police to gauge the impact of crime in the region you are considering.

Size and price

An intensive micro-herb operation or mushroom farm can be profitable on just a few hectares of land, but a beef operation will need hundreds or even thousands of hectares. Depending where it's sited, a small parcel of land on the city fringe could be more expensive than hundreds of acres in another location. Consider what you want to farm, how much land you will need and how much money you will need to be left with after the land purchase to make it all happen.

Infrastructure

Having thought about all the above issues, you've decided on an area and now it's down to choosing a farm within the region. It's time to consider soil type, water security and infrastructure, so what's already on the farms you are comparing? Housing, shedding, irrigation, fencing, established trees, improved pastures, dams, roads, yards, stables, greenhouses? What condition are these in? What will it cost to maintain them, build them or replace them? How do the farms compare?

Summing up, when investigating a potential location for your farm, do plenty of research so you know what you're getting yourself into. Then decide if this really is the best location for your planned operation, or whether a farm an hour or even a state away might be more suitable.

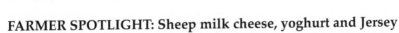

FARMER SPOTLIGHT: Sheep milk cheese, yoghurt and Jersey milk

A builder and his family escape the city, nailing a new lifestyle with a gold medal for their halloumi at the Sydney Royal Cheese & Dairy Produce Show

Farmer: Ian and Jill McKittrick
Farming: East Friesian (European) Awassi (fat-tailed Middle Eastern breed) dairy sheep and Jersey cows
Farm name: Ewetopia
Farm size: 53 ha (130 acres)
Average annual rainfall: 1 200 mm
Where: Ellenborough, mid-north coast, NSW
Farming since: 2002

The McKittricks left Sydney in 2001 knowing they wanted to raise their children in the country. They spent a year living and working in Port Macquarie, taking time to scope outlying regions for land where they could start an agricultural venture.

"We bought the farm based on the worst farming decision," laughs Ian, "its amazing view! We didn't check the soil and we didn't check that the green stuff wasn't weeds, but we don't regret it at all. The land works well for what we're doing because sheep are better off in challenging country on the coast. Also, we've really enjoyed the part of the journey that is finding out what works," said Ian.

After buying the farm, the McKittricks sought to farm something unique in the area. They attended cheesemaking courses and started focussing on sheep milk. They began slowly building the business around off-farm income opportunities and now produce sheep milk fetta, halloumi, labne and yoghurt, as well as Jersey milk as a sideline.

At the time of writing, Ian was milking 23 ewes a day, but he has a vision for a herd of 100. "Rome wasn't built in a day," he said, "but you do need to lay a brick every now and then."

One of those bricks was laid in 2012 when he built a three bedroom farmstay cottage to add an additional income stream and provide opportunities to connect with people and customers. They've also formalised processes when it comes to key farming dates such as when the sheep will lamb, when the ram goes in and when the sheep are to be dried off. As the business is quite seasonal, they structure family life around the farm. The kids tend to play winter rather than summer sport, and maintenance jobs and tractor servicing and the like are scheduled for winter rather than the busy summer months.

"It's a juggle," said Ian. "When you work for yourself there is time and money and you generally don't get both at the same time – if you're busy you have the money, if you're not so busy you'll have the time."

Working 60+ hours per week (Ian) and 30 hours per week (Jill) (80% on physical labour and 20% on the business for Ian and the reverse for Jill) involves twice daily milking (two hours in the morning and two hours at night), feeding lambs and moving them in electric netting twice a week, checking water troughs, processing milk and two days of cheesemaking and packaging. Ewetopia sells direct and via farmers' markets.

"We get great satisfaction from producing a product from beginning to end," said Ian. "I don't think we could do all the hard work of farming and then just have a truck come and pick up the milk, because so much enjoyment comes from interaction with customers."

"And so much enjoyment comes from living and working together," said Jill. "You can't beat that or the sunsets."

Food for thought from Ian and Jill

What are the basics required for your farming activity?
You need Council and Food Authority approvals for cheesemaking and you will need to put together a food manual (ours ran to a couple of hundred pages) including hygiene and handling procedures and records of all farm inputs such as feed, medicines and paddock applications. The manual sets out procedures for dairy operations, maintaining, cleaning, washing up, renewing/replacing milking lines right through to production, processing, storing and market transport and procedures. It needs to list absolutely everything. You can't just say 'wash up'; you need to specify the exact procedure, such as 'fill sink with water, cold rinse first, wash in hot water with detergent' etc....you even need the product sheet for the detergent. It's also important to attend cheesemaking courses and learn how others do it. Attending a TAFE course can help you build networks. You'll go on field days and can also learn other skills such as welding that are invaluable.

What are the threats to the industry?
Supermarkets dictate terms because for the most part the industry is not producing a product, it's producing a commodity with a short shelf life. There are also signs of consumers turning away from dairy due to issues with animal husbandry, environmental and dietary concerns.

Opportunities in the industry?
People are looking for alternatives to cows' milk products, and sheep milk offers that. Customers are also interested in artisan products sold directly to them by local farmers.

What makes a good farmer?
Someone who perseveres, is open minded when it comes to trying new things and has business sense including staying on top of cash flow, being wise when buying and selling things and being aware of economies of scale. Being practically minded and able to fix equipment

and buildings means you don't need to wait for and pay tradespeople, which can mean huge long-term savings. Being a good communicator encourages helpful rather than antagonistic relationships with food inspectors.

Best piece of advice from another farmer?
Just because something has worked in the past or now, doesn't mean it will in the future. Be prepared to adapt.

Worst piece of advice from another farmer?
That won't work. You can't do that. We tried that, it didn't work.

What would you farm if you weren't involved in your current activity?
Not another animal because of the emotional turmoil when they die. Probably some type of plant or a product with a long shelf life.

Ian and Jill's top 5 tips
1. At some point you just need to give it a go and get started.
2. Trial things on a small scale first. We brought home just eight sheep to see if we could do it and liked doing it. Only after the trial did we scale up.
3. To be able to dictate price, small farmers need to produce products, not commodities, but first make sure there's a market for your product. There can only be so many farms in a region doing the same thing.
4. Consider your packaging and plan it well in advance to avoid delays.
5. Start small, get your food safety and product right, and then enter a show or award for the potential publicity and promotion.

What's the best way to enjoy your product?
We love the leftover blue-cheese tasters at the end of a farmers' market. They're a little warm, and are great on baguettes. We're not really into fancy stuff. Just fry halloumi in a pan with olive oil, or enjoy just eating the cheeses straight – when it tastes so good there's no point putting it on a cracker!

7

FARMING PRACTICES AND PHILOSOPHIES

I f you can already spout agricultural terms, definitions and philosophies faster than a 72 metre centre-pivot irrigation machine spouts water, then feel free to skip this chapter and head to the next one. But, if you have a thirst for knowing more about the various branches and styles of agriculture, here's hoping this section waters your curiosity and fills your brain dam with terminology, ideas and concepts.

Agriculture

Growing alfalfa, tending Saddlebacks, shearing Merinos? That's agriculture. Raising buffaloes, trellising passionfruit, plugging branches with shiitake spawn? That's agriculture. Agriculture is basically the breeding and raising of plants, animals and fungi by humans. Within the overarching field of agriculture are various specialties and philosophies.

Horticulture

Horticulture unites green thumbs, dirty thumbs and all thumbs. It's the oxygen-pumping, chlorophyll-laden branch (pun intended) of agriculture. Horticulture is about all things plant.

It's the art behind the moss-smooth bowling green, the spectacular garden clinging to the side of the 40 storey building and the plant propagation required for the fancy garden designs featured in home improvement shows.

It's the science of companion planting and pest control, molecular mechanisms and genomics.

It's the business of cultivating plants to feed, clothe, medicate, decorate and fuel humanity.

Growing echinacea, lemon balm and tulsi for tea? That's horticulture. Regenerating landscapes with shrubs, trees and groundcovers? That's horticulture. Growing pineapples, pecans or *padina fraseri* (a seaweed)? That's horticulture too.

There are numerous specific areas of interest within horticulture and descriptions of just some of these follow.

Viticulture

Grape growing is so specialised and the combined learnings of humanity are so large that viticulture commands a whole branch within horticulture.

If you've ever read the tasting notes on the back of wine bottles or bar menus, you might have come across descriptions such as 'with elements of wood smoke, oak and mandarin, and the perfect seesaw balance of acid and tannin, one sip and you'll arrive at an adult's playground of quaffability,' or 'notes of cherry, ripe plum and pomegranate slide across the tongue followed by a hint of cinnamon, a hit of myrrh and a hot halo of pepper-berry.'

Well, if you plan to harvest Cabernet, Cienna or Chambourcin, you'll need to understand less drinkable descriptive terms such as bot canker, downy mildew and Eutupa dieback. Welcome to the world of viticulture, the world of growing and harvesting grapes. Welcome to a world where the science of grape selection, soils, temperature, pests and diseases can be managed to improve grape yield, characteristics and quality.

For people seriously into swirling, mouthfeel and finishing notes, there is another specialist field called oenology, which is the science and study of winemaking.

Floriculture

Fields of bee-kissed lavender, huge smiling sunflowers and packets of delphinium seeds, perfume bottles brimming with scents of jasmine, lilies and violets. Potted pink peonies, soothing chamomile oil and roses bunched for Valentine's Day. All these things exist in commercial quantities thanks to floriculture, the science of creating and growing new plants from seeds, cuttings and bulbs.

From carefully breeding new cultivars (for colour, form and hardiness) to managing pests and diseases while the plants grow in pots, beds, fields, glasshouses and greenhouses, floriculture also involves postharvest storage right through to marketing and selling.

Turf management

If you literally like watching the grass grow, then this is a specialised area within horticulture that enables just that. The orchestra of weekend lawnmowers and the green grass found at countless footy grounds, cricket pitches, golf courses, bowling greens, schools and new housing developments are all reliant on this form of intensive plant agriculture and after-care. Turf is rolled out for erosion control, landscaping aesthetics and play, and a lot of science goes into creating the right grass for different uses and climatic conditions.

Landscape horticulture

The field of landscape horticulture is about the growing and maintenance of plants for landscaping, as opposed to plants that are grown for eating. It is also about the science and art of designing aesthetically pleasing as well as functional gardens. Landscape horticulture involves everything from production nursery work through to maintenance of parks and gardens.

Olericulture

Imagine a world without the smell of sizzling onion, the comfort of potato, the summer quench of watermelon and cucumber. Imagine never again hearing the snap of a sugar pea, the pop of some corn, the crunch of a carrot. Imagine your taste buds shrivelling up when they never again get to savour the sweet acid of tomato, the sauerkrauted cabbage, the punch of a radish. This would be a world without olericulture, and an unappetising world at that.

Olericulture refers to the breeding, growing, harvesting, storage, processing and selling of non-woody, edible plants, that is, what most of us refer to as vegetables. From iceberg lettuce to chilli peppers, curly kale to straight-laced asparagus, olericulture helps nourish the world.

Pomology

Pomology is not the practice and science of knitting beanies with perfect pompoms on top. It is the knitting together of everything fruit and nut trees need to produce food for humanity (and hungry fruit bats). Mostly focussed on tree crops, it's about everything that these plants need to grow and produce, including soil amendments, pest control, grafting and pruning right through to breeding new cultivars for particular climates and conditions.

A focus on pomology has given the world cultivars such as Australia's Granny Smith apple (a fluke of nature that was discovered and then further developed by a multitude of growers) and the Pink Lady apple through to overseas cultivars such as the sweet and juicy Washington Navel orange.

Pomology in the macadamia industry has led to cultivars being bred for different bearing times, oil content of the kernel, disease resistance, shell thickness and nut fall at maturity.

Arboriculture

If Tarzan were alive today, because he liked to spend so much of his time in trees, shrubs and vines, he'd probably work in arboriculture. This is the science and practice of growing and caring for perennial woody plants. How does arboriculture relate to farming? Perhaps you're thinking of putting in a Christmas tree or oak plantation. Perhaps you're planting a large windbreak and shelter belt, or need to prune or fell trees for timber. Maybe you want to grow the fence posts you will need 15 years down the track. Whatever the reason, you will be able to turn to this sector of horticulture for answers.

Trophort

There's a ukulele playing, the fragrance of ripe mango and a straw poked through the shell of the firm, round coconut. There are blackish purple acai berries, Atherton raspberries and tamarind pods, and what's that? Latex dripping from a rubber tree? Yes, readers, we're in the tropics. Trophort, or tropical horticulture, focuses specifically on plants that need tropical or subtropical conditions to thrive. These conditions can be found in nature, or in greenhouses.

Aquaculture

Those sushi trains chugging around major cities and urban shopping centres are likely hauling farmed tuna and salmon rather than that fished on the open sea and farmed seaweed rather than foraged.

The rainbow trout, silver perch and marron on your plate probably never swam in a river in their lives, and the oysters, sea cucumber and mussels probably had a farmer looking after them too.

Whereas commercial fishing relates to the catching of fish in the wild, aquaculture is the planned farming, in fresh or salt water, of aquatic species ranging from algae and seaweed to molluscs, crustaceans and fish. Aquaculture involves the science and practice behind offshore, estuarine or land-based farms and the species they are being used to grow.

Hydroponics and aquaponics

The Atlantis of farming, hydroponics is the growing of plants in water rather than soil. The ruby red truss tomatoes you see in supermarkets have probably been grown hydroponically, as most likely have many of the strawberries, herbs, cut flowers (such as carnations, roses and gerberas), lettuce (varieties such as butter, cos, mignonette and oak leaf), eggplant, capsicums and cucumbers. A well-run 1.4 ha (3.5 acre) aquaponic farm can produce around a million heads of lettuce per year.

Hydroponic operations can be inside, such as in glasshouses, greenhouses and poly tunnels, or out in the open or under shade.

Aquaponics introduces fish, crustaceans and other aquatic animals into the system in a symbiotic relationship whereby the plants help to filter the recirculating water and the aquatic animals provide the nutrients the plants need to grow.

Setting up these types of operations requires significant expenditure on infrastructure, sound technical skills and a good understanding of horticulture.

∼

FARMER SPOTLIGHT: Barramundi, hydroponic lettuce and herbs

On flat land protected by sand hills, an electrician turned farmer grows barramundi in a recirculating aquaculture system (RAS), and then uses the enriched water from the operation to grow not just lettuce and herbs, but his reputation, winning NSW Farmer of the Year 2016.

Farmer: Nick Arena
Farming: Barramundi, hydroponic lettuce and herbs
Farm names: Tailor Made Fish Farms
Farm size: 17.4 ha (43 acres)
Where: Bobs Farm, Hunter Region, NSW
Average annual rainfall: 1 300 mm
Farming since: 1997

An electrician and builder by trade, Nick Arena was doing home kitchen and bathroom renovations for a couple of clients who happened to be marine biologists. Over the course of the renovations, talk turned from laminates and tiles to how the oceans were being fished beyond sustainable limits.

"It opened my eyes to the impact of the increasing global population and demand for fish, and the potential for devastation of fish stocks. I discussed the idea with family and friends and was offered investment funds (approximately $700,000 in 1997, equivalent to about $2.5 million today) to get an intensive land-based fish production farm up and running."

Nick was determined to farm fish and hydroponic lettuce, and chose barramundi because of the species' quick-growing nature and extensive market appeal.

"We were the first farm to be issued a license to farm barramundi on the eastern seaboard in NSW," said Nick, and as of writing, they remain the only one. In 2010, the business added an onsite barramundi restaurant at the farm.

Nick's typical 60+ hour week (30% physical labour, 40% admin/management and 30% restaurant participation) involves repairs, maintenance and fish harvesting alongside mentoring staff, production planning, the financials, consultancy and technology sales for the business. The products are sold through a mix of wholesalers, retailers, farmers' markets and at the farm gate and restaurant. The business also exports the technology they have developed for fish farming to farmers overseas.

Being at the forefront of the industry meant lots of experimentation and design of systems in the early years.

"It was a difficult start for the business," acknowledges Nick. "There was a lack of expertise and no purpose-built equipment for what we were trying to achieve, but this led to us implementing some of the best solutions including an efficient filtration system to maintain the environment and a passive solar heating system enabling free heating of water to achieve the ideal temperature for the tropical species we grow."

Farming is much harder work than Nick ever expected, and he acknowledges there are longer hours and a lower income in farming, but the benefits have included allowing his children to grow up in a less-crowded and closer community, and he revels in the shortest of commutes.

"The worst thing about farming barra is when an alarm goes off at 2 am because a pump has failed, but that's quickly forgotten thanks to the challenge and personal satisfaction that comes with being able to consistently produce a good product."

Food for thought from Nick

What are the basics required for your farming activity?
Secure water licenses and a government-issued license to farm fish.
Coastal land is more temperate and offers good access to major markets.
You also need special equipment, training and support from someone
who knows what they are doing. Trade skills in plumbing and electrical
are an advantage, or you need to be a great problem solver. You also need
solid commitment.

What are the threats to the industry?
Increasing costs of production such as labour, electricity and fish food,
and the Australian dollar making imported products cheaper.

Opportunities in the industry?
People want value added products, and I see a future for more direct
online sales. I also think farming of the future needs to have what I call
the three golden rules: control, control, control. Being able to control
environmental conditions will ensure more consistent and productive
results that equate to being more viable. Producing more product from
less space with fewer people is part of this philosophy, as in every country, land and labour are becoming more expensive and younger generations are not wanting to continue in the family tradition or business of
farming.

What makes a good farmer?
An observant, hard worker with broad practical knowledge in several
trades. Some marketing and sales skills can certainly assist with getting a
better price for your product.

Best piece of advice from another farmer?
A marine biologist encouraged me to see that fish production needs to
be done in an environmentally friendly and sustainable fashion.

Worst piece of advice from another farmer?
It was actually from another marine biologist who understood fish and
environmental factors, but lacked practical knowledge on establishing
and running an intensive aquaculture system. He just couldn't give me
advice that transferred between the two disciplines.

What would you farm if you weren't involved in your current activity?
If I wasn't farming I'd be working back in my trade.

Nick's top 5 tips
1. Do your homework in relation to what market you will be supplying.
2. Understand the expected sale price for your product.
3. Get good advice from a farmer who is actively doing what you are intending to do, not from an equipment salesperson.
4. Ensure you make good allowances when preparing a budget for establishing your project.
5. Ensure you have sufficient funds to support the project and have a good relationship with your banker.

What's the best way to enjoy your product?
Barramundi is versatile; just bake or grill with lemon pepper butter or with a simple light batter, aka fish and chips.

~

Apiculture

Apiculture is the name for the knowledge and practice involved in the keeping of hardworking, nectar of the gods-producing honey bees. It involves attending to hive care, the feeding needs of the colony and breeding, as well as disease and pest prevention needed to maintain a hive for pollination purposes or honey, wax and propolis production.

Pastoralists/graziers

Pastoralism is the management of domesticated livestock on a farm, normally a large farm or station where efforts are concentrated on fattening sheep or cattle for market.

Mixed farming

Mixed farming can occur on small and large farms, and occurs when the farm grows crops as well as rears animals. Typical examples are wheat and sheep farms, piggeries and agroforestry, and cattle farming when lucerne and/or oats are also grown as fodder.

FARMER SPOTLIGHT: Cattle, hay and silage

A fifth-generation farmer farms the land he was raised on, but keeps reinventing how he manages the land and runs the business.

Farmer: Jock Cuninghame
Farming: Santa Gertrudis cattle, plus hay and silage making
Farm names: Hallcraig, Kilmaurs and Lower Four Mile
Farm size: 1 093 ha (2 700 acres)
Where: Deepwater, NSW (New England Tablelands)
Average annual rainfall: 790 mm
Farming since: he was born

Jock grew up on the land when it was the done thing to leave school and return to help out the family. He eventually became a partner in the operation, then started running it after 15 years. His own son has joined him back on the land now too, and they haven't been afraid to reinvent their farm practices to keep up with agricultural advances. Though their farm is not technically small, it's a traditional small, family-run operation.

"I guess I started out doing what my father had done, but I've gradually branched out," Jock said. Innovations included selling all the ground cultivation machinery (ploughs, etc.) and moving to direct drilling of seed. He also began breeding Santa Gertrudis cattle, a large, hardy beef breed. "Our pastures have also been going gangbusters since we started using feedlot manure as fertiliser."

Some of the key traits Jock thinks are needed to be a good farmer include patience, humour, the ability to be a jack of all trades and humility. "When things turn pear-shaped – and they often do – you need to be calm and collected, and be willing to listen and learn rather than be a know-it-all. Most country people dislike big-noters."

"I sometimes think you have to be born into a rural lifestyle to really appreciate it and understand it," said Jock. "You have to realise that work doesn't stop at 4 pm on Friday afternoon and start again on Monday morning – and that there's no quick and easy way of making a dollar. You have to just keep grinding away, doing the same stuff over and over, but lots of little bricks can make a very impressive wall in the end."

Jock's typical 60+ hour week (85% physical labour and 15% admin) involves fencing, all types of cattle work, farm maintenance and tractor work including spraying, seeding, and hay and silage making. He sells wholesale and retail and makes silage and hay to ensure feed certainty for his stock in times of drought and as a paid service for other farmers.

"Animal care knows no holidays, so you can pretty much be tied to the place," says Jock, "but the peace and solitude of country living and not having to take orders from anyone make up for it. There's no one around to annoy you, no traffic jams, and there's so much satisfaction seeing fat shiny animals standing gut deep in lovely green tucker."

"It is a lot harder to make a living from the land than everyone says, especially if you need to borrow heavily to get into the industry. But if you love what you are doing and have a real passion for it, the lifestyle farming brings you is far more important than being the richest but most miserable person around."

Food for thought from Jock

What are the basics required for your farming activity?
Good fertile country, water security and good business sense. You also need cattle that really perform for the land and climate you're in and good machinery to get the job done...and if it breaks down, you need to have your handyman skills handy.

What are the threats to the industry?
The high costs and time demands of farm maintenance and infrastructure. There's also competition from other countries in relation

to beef exports, and there's always the challenge of keeping Australia disease free.

Opportunities in the industry?
The beef industry goes through ups and downs, but currently the opportunities are pretty good – you just need to keep producing a good article that is readily saleable.

What makes a good farmer?
A hard, but smart worker who uses the brain as well as the back. An easygoing character who is trustworthy and can be relied on by others.

Best piece of advice from another farmer?
Pick other farmers' brains.

Worst piece of advice from another farmer?
Stick at it, you'll be a rich man one day.

What would you farm if you weren't involved in your current activity?
Definitely not sheep, I did 20 years of that! We can have terribly cold winters here, so I'd like a mango farm somewhere warm. I would love to sit under my mango trees in Mareeba and eat mangoes until I get the scours!

Jock's top 5 tips
1. Never be afraid to ask someone who knows more than you.
2. Back your own ideas and judgement.
3. A farm is a business. You need more than just a strong back and arms.
4. To make money farming, you must spend big money. Get used to that idea.
5. Have some off-farm income to hedge your bets.

What's the best way to enjoy your product?
You can't beat a really good BBQ steak. Use the good cuts (sirloin, fillet, ribeye, rump). Find one that has some intramuscular fat for flavour and tenderness. Cook it medium rare and wash it down with an icy cold beer

in a glass that has been in the freezer! Life does not get a lot better than that!

Diversified farming

An extension to mixed farming, diversified farms grow a variety of plants and potentially rear multiple species of livestock. In the case of smaller farms, it might well mean a VERY mixed farm where the landholder produces everything from eggs, honey and salad vegetables to beef, oranges and lemons. The aim is for these enterprises to work as a system where the by-products of each activity (e.g. manure and waste oranges) enrich and support the others (e.g. soil fertility and cattle feed), thereby adding to the viability of the operation.

Broadacre farming

Broadacre farming refers to large-scale cropping, mixed cropping and large livestock operations. If you're imagining combine harvesters as big as houses, GPS-directed fertiliser programs, irrigation channels and cattle yards the size of a small town, you're on the right track.

According to the Australian Bureau of Agricultural and Resource Economics and Sciences (ABARES), these farms are responsible for managing more than 90% of the total area of agricultural land in Australia and account for the majority of Australia's family-owned and operated farms. Located in all regions across Australia, these farms form a vital part of rural communities and local economies.

Broadacre farms aren't your average coastal farm in size; they are set up on huge swathes of land and are capital intensive. They require significant capital investment in plant such as machinery, and might grow oilseeds, cereals, sugarcane, lupins, rice, hay and silage, as well as fibre plants such as cotton and industrial hemp.

The specific types of plants grown might include barley, canola, chickpeas, faba beans, oats, wheat, peanuts, lucerne, safflower, sorghum, maize, hops, sunflowers, cotton and mung beans. The types of livestock run on broad acre farms are cattle and sheep...lots of them!

These farms can operate conventionally through ploughing, or they might be no till or minimum till, meaning that the seeds are direct drilled through the residue left by the previous crop.

Dry farming

Dry farming doesn't mean you can't enjoy a tipple, it just means that the tipple must come from rainfall rather than irrigation. Dry farming relies on good soils soaking up moisture in the wet season to release it when it's needed in the growing season. Special techniques are used in dry farming to make the most of the available moisture. These techniques range from terracing and wider spacing between plantings to creation of swales and zero tolerance of weeds that would otherwise suck out the available moisture.

Irrigated farming

Let's go back...way back...all the way back to when Egyptians were settling into their new digs near the Nile and some bright spark decided to divert some of the floodwaters to the fields. A few hundred years later, King Menes and those who came after him experimented with dams and canals and created a lake. By 700 BC there were waterwheels, and farmers used gravity to channel water downhill from springs, and hundreds of years later, windmills, tanks and pumps (hand-pumped, fossil-fuelled, electrified and now solar) shunt water sideways, upwards and downwards to wherever it's needed.

Irrigated farming is any type of farming that relies on water that's been captured or channelled, bored for or obtained from springs. It might be piped, hosed, sprinkled, dripped or flooded, and it brings life to crops when rainfall is insufficient or when yields need to be improved.

Intensive farming

Intensive farming is a method of farming that attempts to increase yield and efficiencies over what would normally be possible from a certain land area. Typical of this method is a high concentration of livestock and/or plants, combined with increased inputs such as capital (often in the form of equipment), labour, pesticides, herbicides, fungicides, fertilisers and medication. These finely tuned inputs can lead to faster growth rates than those experienced in traditional agriculture.

Planting an acre of sunflowers by hand? That might feel intensive on your back, but it's not intensive agriculture. Intensive agriculture is when systems and equipment are used to do things like the following.

- Plant five million sandalwood trees across 2 200 ha in Western Australia.
- Produce 350 tonnes of hydroponic tomatoes a week out of 20 ha of adjoining greenhouses in the South Australian desert.
- Collect 50 million litres of milk a year from 3 700 dairy cows living across 2 700 hectares.
- Feed and manage 26 000 cattle on an 800 ha (1 977 acre) feedlot.
- Raise 40 000 adult meat chickens in a 150 m long by 15 m wide shed, oh, and there are eight other similar sheds on the property.
- Welcome 500 piglets a week into the world.

Okay, that's probably the upper end of intensive farming in Australia, but intensive farming can also mean millions of tiny seedlings growing in polytunnels; one acre of hydroponic lettuce; 200 acres planted with summer and winter crops or 100 000 banana trees on 100 acres.

Organic farming

Customers, arms straining from bags overflowing with produce, reach for their wallets one more time as they walk under the sign 'Don't panic, it's organic'. It's at city and large regional farmers' markets where customers swoop early to get the best of the fresh produce. Organics are packaged not just within the skin of fresh fruit and vegetables, but in cryovac bags of nuts and meats, in yoghurts and bags of chips, and in everything from rice to sugar.

The amount of organic produce available to Australians and New Zealanders continues to increase, although not all of it is locally grown.

A quick look through a supermarket freezer might reveal organic berries from Chile or China alongside organic peas from Europe.

So, what is the definition of organic farming? It depends on who you talk to, but according to Organics International, and the emphasis is theirs, it is *"A production system that sustains the health of soils, ecosystems and people. It relies on ecological processes, biodiversity and cycles adapted to local conditions, rather than the use of inputs with adverse effects. Organic Agriculture combines tradition, innovation and science to benefit the shared environment and promote fair relationships and a good quality of life for all involved."*

What does that mean on the ground at a farm? It means things like:

- improving the soil by making compost and growing nitrogen-fixing crops (green manures) rather than using synthetic fertilisers
- planting and protecting habitats for predator bugs
- minimising weeds by using mulches, tools and, when necessary, flame or steam weeders
- not using GMO seeds
- creating buffer zones to avoid contamination from roads and neighbouring conventional farms
- not using growth promotants, synthetic pesticides, herbicides, fungicides, drenches, vaccinations and pour-on treatments.

Farms can choose to operate organically, but if they want to be certified by one of the accredited organic certifiers in Australia they are required to go through audits and conversion periods and adhere to strict rules. The following information has been provided by Australian Certified Organic (ACO), Australia's largest certification body and owner of the most recognised organic logo amongst consumers – the ACO Bud logo. ACO provides domestic and international certification services, including those relating to organic production, non-GMO plants and food safety.

What is ACO's simple definition of organic farming?
Organic production systems are guided by the following principles and outcomes: production of naturally safe, high-quality, nutritionally vital

foods; optimal production output, with rational and minimised use of inputs; use of recycling and biological cycles within the farming system; biodiversity protection and enhancement within the farm and surrounding areas; regeneration of lands and soils and best environmental practice of farming activities (as per the Australian Certified Organic Standard).

How long does it take to become certified?
A processor/handler can become certified within 6–8 weeks of applying for certification, so long as they are approved as compliant with the Australian Certified Organic Standard by an onsite auditor and the ACO Certification Team. A producer can take between one and three years to become fully certified organic, depending on their previous level of organic compliance. If a producer can provide records of organic compliance for a minimum of 12 months prior to applying for certification, they may skip the pre-certification period and be granted In Conversion status after the first successful audit. If a producer cannot prove organic compliance for the previous three years before applying for certification, they will receive Pre-Certification status for a minimum of 12 months followed by In Conversion status for 1–2 years.

What are some of the things farmers must do to become certified?
Apply and submit documents outlining the management of the operation.

What are some tips and advice you would give to farmers thinking about becoming organic?
Keep up-to-date records and use inputs that are allowed for organic production and processing.

What are some of the things that are banned under organic farming?
In addition to synthetic chemicals, organic produce and inputs into organic produce may not contain genetically modified organisms, may not be treated with ionising radiation, and cannot use nanotechnology.

What happens if an animal needs antibiotics/pharmaceutical medi-

cine to save its life? Can a farmer use the medicine and then withhold milk/meat for a period, or is any medicine use banned?

The Australian Certified Organic Standard 2016, sections 5.1.1 to 5.1.6, notes that practical measures shall be maintained to ensure livestock health. Vaccines shall be restricted to those used for a specific disease that is known to exist in a region and threatens livestock health and those that are required by law. Use of prohibited treatments require written veterinary advice and shall lead to decertification of stock and will require stock to be quarantined for three times the legal withholding period or three weeks (whichever is longer). Organic re-certification of livestock products may take place as follows: wool – 18 months after treatment, milk – 180 days after treatment, eggs – 60 days after treatment. Meat has permanent loss of organic status.

FARMER SPOTLIGHT: Organic sultanas and citrus

A newspaper photographer and her two daughters trade Melbourne for the Mallee.

Farmer: Erin Jonasson
Farming: Sultanas and citrus
Farm name: Sultana Girl
Farm size: 20 ha (50 acres)
Where: Goodnight, NSW (edge of the Mallee and Murray River region)
Average annual rainfall: 300 mm
Farming since: 2009 (full-time since December 2010)

Erin spent 20 years snapping photos for the Australian Financial Review, and then one day snapped her trigger finger. Years of wear and tear lugging heavy cameras and years of wear and tear living the newspaper life finally saw her step into a frame of her own.

"I'd spent a lot of time out in the Riverina area photographing and interviewing farmers, and I just thought it looked like the nicest life in the world," said Erin. "Once I decided it was what I wanted to do, I looked at more than 20 farms, including bare land. I didn't want to grow vegies or animals, so I focused on horticulture, which is basically permanent plantings. Financially, I couldn't wait years for the first crop to come in, so I opted for a farm with all the infrastructure in place and vines already in the ground."

Drought at the time meant that prices were affordable, and the elderly farmer she purchased from stayed on in the area and acted as a sounding block and invaluable mentor. Erin converted the farm to organic, and then in 2010 suffered a major setback with massive crop losses from flooding.

"That's when I learned how important diversity on a farm is, as the citrus helped get us through. We value added everything, went direct to farmers' markets for the best price, juiced them and generally got as much out of the citrus as we could to make up for the sultana losses."

Erin's typical 60+ hour week (80% physical labour and 20% marketing/admin) involves slashing, spraying, packing sultanas, picking oranges, sending out orders, following up retail orders to keep stock topped up, irrigating and fertigating. She sells her product online and via wholesale, retail and farmers' markets, and exports to Vietnam. She takes time off in winter when the vines are dormant, and once a month drives 400 km to Melbourne for a farmers' market and to catch up with friends, see shows and enjoy the city she left behind. Then it's back to the farm she loves.

"My kids came home from an economics lesson at school where they'd learned about primary (producers), secondary (manufacturers) and tertiary (retailers) industries. They said, 'you do all three, Mum'. That's one of the things I really love about farming – we can go from creating right through to selling – so our brain works across all three levels. It's challenging and fun, and I never get bored. There's no reason we city people can't learn to farm."

Food for thought from Erin

What are the basics needed for sultana farming?
There's so much information out there that it can be confusing, so people need to get on the phone and/or go and meet other sultana growers, that is, talk to the people who are doing it. You'll also need access to labour during harvest time, and land of course, but you don't have to go

too big. Niche farms with quality products can be economically viable. If I'd bought or planted another 20 acres, I'd have to employ a full-time assistant.

What are the threats to the industry?
Large agribusiness can turn a fraction of their land organic, undercut the market and, in the process, squeeze out smaller operators, and dodgy operators can hurt the trust in organics.

Opportunities in the industry?
There's a huge consumer push for more organic food. Having your own website will help buyers, retailers and exporters to find you.

What makes a good farmer?
Someone who really spends time keeping an eye on whatever it is they're growing. It's about being led by what you observe and your intuition, not by schedules and textbook farming. Farmers also need dedication, self-belief and flexibility, and to understand that trends can come and go, so it's good to stick to your own ideas and find a market.

Best piece of advice from another farmer.
When Erin was freaking out about getting things wrong, another farmer helped take the pressure off by saying, "What are you going to do? Sack yourself?"

Worst piece of advice from another farmer.
I'll show you how to use Roundup.

What would you farm if you weren't involved in your current activity?
They have a huge set-up cost, but probably pistachios.

Erin's top 5 tips
1. Find new markets, don't just stick with traditional markets or wholesalers.
2. Reach out to the public.
3. Use a jack rather than your muscles.
4. Don't be scared to go organic.

5. Brand your product well. You don't have to name it after your region; choose a name that will resonate with the public.

What's the best way to enjoy your product?
Sweet sultanas with a rich, dark chocolate and salty nuts!

∽

Conventional farming

What is now termed 'conventional' farming only became the predominant farming method in modern agriculture after World War II. The bulk of food, fibre and flower production in Australia and New Zealand comes from farms using this approach.

Also known as industrial farming, these farms lean toward the use of inputs such as synthetic chemical fertilisers, pesticides, herbicides and fungicides. The farms make use of technology and modern machinery and tend towards monocultures or limited polycultures. They might potentially use GMO feed and seed.

A conventionally run farm might spray glyphosate on weeds, use broad spectrum drenches on sheep, and fertilise using petrochemical derived fertilisers such as Ammonium Nitrate and Superphosphate. The same farm though might also make use of biological tools such as releasing good predator bugs, administering natural feed supplements and making and spreading compost.

Biodynamic farming

Since autumn, a cow horn packed with fresh manure, has lain buried 40–60 cm below the surface in rich soil. It's now spring, and it's being carefully dug up. It needs to be rhythmically stirred for an hour in rainwater before being sprinkled across the farm in the late afternoon at a rate of 85 g per hectare. The aim of this practice and formula, known as Biodynamic Preparation 500, is to activate the soil to improve germination, root development, plant growth, worm activity and humus formation. This is just one biodynamic practice that was developed by Austrian Rudolf Steiner. To some, it sounds like woo woo, to many others biodynamics forms the framework needed for a successful and ethical farm and life.

Permaculture

Australia has been responsible for many ground-breaking agricultural advances including mechanical shearing clippers, self-propelling rotary hoes and utes. Another invention came about in 1974 when a collaboration between Bill Mollison and David Holmgren led to permaculture, an approach to farming they hoped would integrate ecological and environmental design to ensure a sustainable form of agriculture.

David Holmgren defines permaculture as "*Consciously designed landscapes which mimic the patterns and relationships found in nature, while yielding an abundance of food, fibre and energy for provision of local needs.*"

A permaculture approach involves making farming decisions around a central core of ethics. These are:

- care of the earth (soil, forest and water)
- care of people (self, kin and community)
- fair share (set limits to consumption and reproduction, and redistribute surplus).

Using these guide posts, the permaculturalist makes the most of the features of the land and the use of energy on the land by mimicking natural systems. There are 12 main permaculture principles that can be applied in different ways depending on the land, climate, needs and people. Thank you to www.permacultureprinciples.com for this information. The principles are:

- observe and interact
- catch and store energy
- obtain a yield
- apply self-regulation and accept feedback
- use and value renewable resources and services
- produce no waste
- design from patterns to details
- integrate rather than segregate

- use small and slow solutions
- use and value diversity
- use edges and value the margin
- creatively use and respond to change.

If you were to visit a permaculture farm, you might see the use of hand tools such as broad forks, food forests brimming with various species, integrated aquaculture, rainwater harvesting and reuse, swale systems and row plantings on contours, and solar passive buildings made of natural materials.

Permaculture gardens and farms are usually small, and profitability is hard to achieve, although it can be done through a mix of earnings from nurseries, teaching others interested in permaculture, consulting, guided tours, produce sales and savings on the costs of energy and food that are eventually produced on site as the property matures.

Biological/regenerative farming

Biological farming is a dirty business. That's because it's all about the soil.

Farming so often concentrates on the big stuff above ground, but healthy animals and plants rely on what's going on below the ground... the beneficial microorganisms and worms that matter so much.

Good soils allow deeper water penetration and offer up crucial nutrients. The six principles of biological farming, as shared by Gary Zimmer, American farmer and world leader in biological farming, are:

1. Test and balance your soils and feed the crop a balanced, supplemented diet.
2. Use fertilisers that do the least damage to soil life and plant roots.
3. Apply pesticides and herbicides responsibly while relying on customised management practices to reach maximum genetic potential.
4. Create maximum plant diversity by using green-manure crops and tight rotations.
5. Manage the decay of organic materials and the balance of soil, air and water.

6. Feed the soil using carbon from compost, green manures, livestock manures and crop residues.

The results, as seen in fields across the country, are the reason why farmers from organic to conventional are taking a closer look at how to integrate biological processes into their farming toolkit.

According to Farmers Without Borders, regenerative agriculture is "Any kind of farming that enables the restorative capacity of the earth. Regenerative agriculture preserves or improves the fertility of the soil, creates an abundance of food and other agricultural products, contributes to vibrant communities and equitable economies, and respects the ecology of the natural world. Fertile soil helps create nourishing food and, in turn, healthy people and robust communities."

Of all the farming philosophies and practices, which ones connect with you?

~

A little about market gardening

Market gardening is about growing a wide variety of seasonal crops, mostly using hand tools and small machinery, on a small area of land. Think well planned plots, beds on contours, well-conditioned soil and a riot of asparagus and artichokes, choy sum and coriander, zebra tomatoes and zucchinis. A market gardener might also use greenhouses and hoop houses to extend the season.

Australia's market gardening heritage owes a lot to immigrants. The Chinese started way back in the 1870's then, according to Australian Government's Department of Immigration and Citizenship, "By the late 1930s, one-third of all Australia's Italian migrants lived in the cane-growing regions of Queensland. Italians also became involved in market gardens, comprising about 40 per cent of Queensland's market gardeners." They were then joined by other waves of immigrants including Greeks, Lebanese, Maltese, Croats, Vietnamese and Africans. It seems young farmers are also increasingly attracted to small scale market gardening and you can read about some specific examples of market gardens in the farmer spotlights on pages 126 and 231.

PART II

8

WHAT TO PRODUCE

G o on, open the treasure chest. Take the bronze key and turn it. Now lift back its heavy carved lid to reveal the opportunities – of all shapes and sizes – glittering within. That's how it felt to us when we started on our own farming journey. It was like stumbling into a treasure room laden with potential wonders. It was so bright at first, we were blinded by choice.

So, we began sifting through the jewels.

Aquamarine? No, fish farming didn't float our boat.

Citrine? Maybe a few lemon trees around the house, but not an orchard.

Emeralds? Instant evocation of flourishing greenery.

Gold? Was that solely what we wanted to go for? Or...

Rubies! The red of passion. We wanted some of that!

So that's what we pursued – multiple passions rather than mono-culture.

The discoveries we made while farming – about nature, ourselves and society – were treasures all on their own, but they did not pay the bills. The bills were paid by selecting key things to farm. For us, that was bees (honey and native), miniature Galloway cattle, turmeric, wild hibiscus and humans (in the shape of tourists). We also farmed multiple other species including alpacas and aniseed myrtle, goats and gotu kola, macadamias and mulberries, but they were part of the framework, not the focus.

So...you want to be a farmer, but what do you want to farm? What will be your focus, your jewel? Will you be led by what you need and want, by what customers need, want and will pay for, or by what your land and capital can most easily support?

There are armadas of opportunities to explore both in and outside Australia's highest-value production areas of cattle, wheat and dairy. Open your mind to going on an agricultural treasure hunt, and don't necessarily settle for what everyone else is doing. Why? Because although at its simplest, farming involves either livestock or plants, at its most complex, it is about producing everything humanity requires for life. That means there is a world of opportunity out there.

This section is all about helping you find your treasure. The lists are not exhaustive, but will hopefully open your eyes to the wide array of farming options producing raw materials, food, beverages, shelter, fibres, fuels, medicines, perfumes, education and entertainment.

Raw materials

Raw materials are the building blocks of everything we use and consume. They are non-processed or minimally processed materials. Oil and coal, as well as sunshine, wind and water are the raw materials of the energy industry, but when it comes to agriculture, raw materials include things such as the following:

- rubber and latex from rubber trees, used in the manufacture of everything from tyres to washing-up gloves
- cork used for wine bottles, in flooring, musical instruments and more
- animal hides for leather (such as shoes and handbags) and floor and furniture coverings
- fibres, the basis of everything from clothes to ropes, including cotton, hemp, wool, silk, and wood pulp for paper
- flowers (cut for food, decoration and gifts, as well as for perfumes, teas and hops)
- dyes such as barberry, madder root, pomegranate peel etc.

- miscellaneous products such as reeds for thatching, loofahs for bathroom use and timber for firewood
- Christmas trees
- birdseed and animal food such as lupins, canola and sunflowers
- plants for landscaping.

Food

Bingo! Humans need to eat. So do the animals that humans eat. So do the pets that humans keep. That sounds like a multi-billion-mouth market. But everyone has different tastes, so let's break it down a little so you can see if anything makes you want a bite.

Culinary herbs and spices

Imagine a pizza without oregano, a summer deprived of basil and an Indian curry missing the flavours of turmeric and cumin. For thousands of years, humans have traded in herbs and spices. Nutmeg, cinnamon, cardamom, ginger...even just reading the words evokes delicious scents and mouth-watering munches. Herbs and spices can be grown in the field or in greenhouses. Some require lots of water, whilst others are drought hardy. Many can be value added.

Grains & seeds

These include wheats (including types such as durum, spelt, freekeh, emmer and einkorn, bulgur), oats, rice, barley, rye, corn, triticale, millet, sorghum, teff, quinoa, farro, kaniwa, buckwheat, sunflower seeds, canola and more.

Vegetables

- Brassicas. Broccoli, cauliflower, mustard, horseradish, wasabi, cabbage.

- Leafy and salad vegetables. Bok choy to silverbeet, kale to kohlrabi.
- Legumes. Alfalfa sprouts to snow pea shoots, lentils to Lima beans, broad beans to kidney beans, peas to peanuts.
- Onion family. Chives, spring onions, shallots, leeks, garlic, onions.
- Root vegetables. Beetroot, carrots, daikon, parsnips.
- Squash family. Pumpkins, zucchini, squash, cucumbers.
- Tubers. What lies below the ground is the potato, sweet potato (leaves, flower, vine), yam, cassava (and its extract tapioca), taro, Jerusalem artichoke, aquatic lotus root, arrowroot (which can also be used as gluten-free thickener).
- When you are a fruit but we call you a vegetable. Members of the capsicum and tomato family contain seeds in the edible section of the plant, which means they are classified as fruits, but most people think of them and eat them as vegetables, which is why we are including them in this section.

Fruits and nuts

- Citrus. Lime, mandarin, orange, lemon, lemonade, blood orange, tangerine, grapefruit, finger lime.
- Pome. Apple, pear, quince.
- Prunus. Cherry, plum, peach, apricot, nectarine. Almonds are also part of this family.
- Oleaceae. Olives.
- Vine fruits. Melons such as watermelon, rockmelon and honeydew, grapes, kiwifruit, passionfruit, dragon fruit, and berries including raspberry, strawberry, blackberry, cranberry, goji, youngberry, keriberry, loganberry, mulberry.
- Tropical fruits. Pineapple, guava, avocado, banana, cherimoya and so many more.
- Nuts. Macadamias, pecans, chestnuts, walnuts, pine nuts, Brazil nuts, pistachios, bunyas.
- Hundreds of other fruit-producing plants ranging from

persimmon and pomegranate to native species including Davidson's plum, muntries and quandong.

Fungus

Shiitake, oyster, button...mushrooms happen to be one of the most valuable horticultural crops produced in Australia. People can't seem to get enough of the fungus family, and they are used in cuisines from Italian to Japanese to your weekend Aussie breakfast. Mushroom farming requires shedding, cool temperatures, cold storage and access to lots of water. Whereas some mushrooms grow on timber, regular field mushrooms like Honey Brown require lots of compost and materials such as sawdust, so access to your farm will need to be suitable for deliveries. There's also the most expensive fungus of them all, the truffle.

Proteins

From grasshoppers, ants and crickets through to snails, fresh and saltwater fish and crustaceans. Rabbit, roo, goat, beef, lamb and buffalo meats. Dairy products and eggs, too.

Beverages

Humans need to eat, and they also need and love to drink! That's why some farming ventures start with a liquid end in mind, while others discover it as an afterthought.

- Animal milks. Camel, sheep, goat, cow and fermented milks such as kefir.
- Plant milks. Almond, soy, coconut, rice, oat.
- Juices. Fruit and vegetable juices.
- Soft drinks. Ginger beer, lemonade, turmeric sodas.
- Teas. Herbal-based blends and kombuchas.
- Chocolate-based drinks.
- Coffee.
- Alcohol. Beer (yes...there is a carrot beer!), wine, spirits, ciders and mead.

FARMER SPOTLIGHT: Certified organic winery

A pharmacist and a horticulturalist meet during a winemaking degree and after years working for other wineries, plant unusual vine varieties to create award-winning wines of their own.

Farmers: Jenny and Eric Semmler
Farming: Grapes including Durif, Tempranillo and Touriga Nacional
Farm name: 919 Wines
Farm size: Two vineyards totalling 40 acres
Where: Glossop and Loxton, Riverland region, South Australia
Average annual rainfall: 243 mm
Farming since: 1999

Jenny and Eric Semmler had worked for a number of impressive wineries but a 2001 trip to the Duoro Valley in Portugal was a game changer that helped them establish their own vineyard.

"The wines we were tasting in Duoro were world class," said Jenny, "and they were growing in a climate just like that in the Riverland, an area three hours out of Adelaide."

Eric, originally a horticulturalist, realised that if they used the right variety of grapes for the climate, rather than just a traditional shiraz, they too could produce benchmark wines.

The couple continued to work off-farm for a few years while setting up the vineyard. They sourced special grape varieties, made their first wine in 2004 and bought a second vineyard in 2011 to ensure supply and continued growth for the company.

Eric had earned his reputation as one of Australia's most respected fortified winemakers when he worked for BRL Hardy/Constellation/Accolade Wines, and brought all that expertise into the business. He is an authority on the production of the heritage Apera and Topaque styles of wine, while Jenny won Best Winery Owner/Operator in the Australian Women in Wine Awards 2016.

Working as many hours a week as needed, the couple's business roles are split with Eric taking care of viticulture and winemaking operations, while Jenny handles logistics, marketing and compliance. They both take care of HR and sales. There is a planned annual cycle of activity involving irrigation, maturity testing, harvesting, crushing, fermenting, pressing and winemaking. Then there is sales work, attending festivals and running the cellar door, and when the leaves fall off pruning starts, which can take several months using a mix of hand and mechanical equipment. Along the way, there are repairs to equipment, logistics for scheduling bottling (for their own wines and other vineyards for whom Eric acts as winemaker), vineyard repairs and replanting vines. September is the start of the Certified Organic spraying program, which allows the use of a restricted set of sprays to ensure a good healthy crop. There is also payroll, the books and marketing to do. The Semmlers sell via restaurants, trade, independent bottleshops, events, the cellar door, online, export and have a distributor.

"We love being in control of own destiny and addressing issues and opportunities quickly. You do need to be part drill sergeant, part gun programmer and part parental hand-holder, and we're continually herding people along to get it all done," said Jenny. "But doing things sustainably, looking after the soil and providing a haven for wildlife gives us a lot of satisfaction."

Food for thought from Jenny and Eric

What are the basics required for your farming activity?
Relevant education such as a Bachelor in Wine Science. You also need hands-on experience and capital, lots of capital. Though wineries sell romance to customers, we're not in a romantic industry. We're in agriculture, and you need a strong business mind. If something's not working, get rid of it or do it differently.

What are the threats to the industry?
Threats come from within the industry, as it's one that attracts people with lots of money to lose. Ego and lifestyle are wrapped around the name on the bottle rather than around the business of running a winery. This can lead to price points being reduced and more pricing pressure being applied to established producers.

Opportunities in the industry?
Fortified wines are making a comeback, but it's an area requiring specialist skills. White spirits are a trend, with opportunities for boutique gin and vodka.

What makes a good farmer?
Someone who looks after their resources and does it with a mix of optimism, persistence and the ability to use practical solutions to solve difficult problems. Also, be prepared to keep learning and changing.

Best piece of advice from another farmer?
To always use off-site events to draw people back for another reason. If we are promoting off-site, we give out vouchers enabling people to visit our cellar door and receive a free tour or glass of wine.

Worst piece of advice from another farmer?
"You'll never sell it at that price." A lot of people in business are frightened to ask for a relevant price for their products, but price can add perceived value to a product. Although if you price a product higher, you need to have earned it through quality and customer engagement.

What would you farm if you weren't involved in your current activity?
Something not everyone is doing, so spices like saffron and vanilla.

Jenny and Eric's top 5 tips
1. It's not always possible, but try and be first and best to market so you are in peoples' minds. If you can't be first, at least be second, but don't be 50th or 200th.
2. Value add, and do things differently to the large producers.
3. Produce different products for different markets.
4. Focus on niche marketing.
5. Be price setters rather than price takers.

What's the best way to enjoy your product?
Good food, good company, good memories, good wine and a gin and tonic with an equal slug of apera (pale dry sherry style), or the Behind Bars Barman Association developed a cocktail recipe called the Apera

Cobbler: two muddled strawberries, crushed ice, a shot of pale dry apera, and garnish with slices of lemon, lime and orange. You get your two serves of fruit a day and your apera with that one!

In addition to food and beverages, there are plenty of other purposes you can farm for, including the following.

Shelter

Between farming and mining, the needs of society in terms of shelter are met. If farming to provide shelter, you might look at the following.

- Timbers for cabinetry, frames, floors and furniture from trees ranging from Radiata Pine to Australian timbers such as Blackbutt, Jarrah, Tallowwood and River Red Gum.
- Thatch for roofs.
- Bamboo for construction and fencing.
- Trees for shade and landscaping.
- Straw for straw-bale houses.

Fuels

Fuel just doesn't come from a bowser at the end of a supply chain from the Middle East. Fuel can also be produced on farms in the following forms.

- Firewood. Kindling to long-burning logs.
- Ethanol. Produced from crops including sugar cane, sorghum, beets, barely, hemp, potatoes, corn, cotton and sunflowers.
- Biofuel. Produced from algae.
- Methane gas. Produced from pig and other animal manures.
- Electricity. Produced by wind, water, solar and thermal power.

Medicines and perfumes

Thousands of years before the rise of big pharmaceutical companies, humans turned to the natural world for medicines. Your farm could become a "farmacy" supplying "nutraceuticals", bushtucker and growing all sorts of healthy offerings.

- Honey including medicinal honey from species such as *Leptospermum scoparium* (one of the tea trees also known as Manuka and Jellybush).
- Chinese medicinal herbs ranging from astragalus to ginseng to wolfberry.
- Common medicinal herbs from chamomile to lemon balm and echinacea.
- You could also grow the foundation plants for modern medicine such as aspirin, or (legal) poppies for opioid production.
- Hemp and medical marijuana.
- Plants to produce oils and perfume ranging from specific roses and irises to lavenders, ylang-ylang and bergamot.

Education and entertainment

These days, a farm can also produce education and entertainment in the form of agritourism, which is covered extensively in Part 5.

Fill a gap in the market

If you're finding it hard to locate a product, or wish someone was producing it, think of it as an opportunity for a business. This is how Neville and Sophia Donovan got started in 2002, and the enterprise they launched is now the thriving family enterprise GreenPatch Organic Seeds & Plants.

"We'd been growing vegies, herbs, rhubarb and fruit to supply the local organic market but kept running into difficulty trying to buy certified organic seeds," said Neville. "That led to us actively collecting vegetable, herbs, fruit and flowers seeds each season and developing an understanding of plant families and what cross-pollinates with each other. We focussed on vegetables and herbs because of the nourishment they provide, flowers because they're pretty and bring an abundance of insects to the garden, and fruit because it's delicious to eat."

Seeing increased demand for organic seeds from others not just themselves, Neville and Sofia decided to completely focus on the seed growing side of the business. They purchased a little over 5 ha of relatively flat land at Glenthorne, just a few minutes off the Pacific Highway and a short drive to the local post office, an operational decision important to running a successful mail order business.

In addition to growing, processing and packing seeds, they produce seed and plant catalogues each year for their mail order business, offer internet sales and an on-farm shop and special events such as the annual Tomato Day where they share growing information and recipes.

"Saving open pollinated, non-hybrid, non-GMO seeds is a passion for us and we get constant affirmation from customers that they appreciate our efforts," said Neville. "Filling a gap in the market has enabled us to grow a business, create jobs for family and other members of the community and do what we love."

In summary, when selecting what produce to farm, you can choose from one or a mix of livestock, cropping, horticulture, viticulture, forestry or fishing. You can choose something you're personally fascinated by, will be proud of, and/or is highly profitable, relatively easy, within your capability, what the market needs etc. You can choose things that you can sell whole, or add value to.

With so many options for your farm, think deeply about your choices. Your crop needs to be something that will sustain you financially, mentally, physically and spiritually in the coming years. It needs to be something that after all the effort and care you lavish on it, will not just earn you a living, but allow for and enhance your life – and that of those around you.

So, will it be fins, fruits or fluff? Morsels, morels or medical marijuana? Hopefully, by the end of this book you'll have a better idea.

9

VALUE ADDING

Value adding is the process of changing or enhancing the raw material you produce in such a way that customers will appreciate it and be prepared to pay more for it.

If you've ever bought trimmed and bundled herbs, sauerkraut, a flower arrangement or beef jerky, you've bought a value added product.

Just look at cow's milk as an example. It can be value added in so many ways including:

- packaged in bottles and cartons
- it comes in full cream, skim and A2, and can be fresh or long-life
- it can be processed into many different types of soft and hard cheese ranging from ricotta to tasty to parmesan
- cheeses can be sold in tubs, blocks, sticks, sliced or grated
- the milk can be powdered
- the milk can be flavoured
- the milk can be condensed
- the milk can be turned into natural and flavoured yoghurts
- it can be processed into butter
- the cream from the milk can be put in cartons or into high-pressure tubes for dessert decorating, or it can be soured

- it can be processed into baby formula and nutritional drinks
- it's a crowd favourite as ice-cream
- and it's used in all sorts of concoctions from pre-packaged alcoholic drinks to body creams and face masks.

Milk provides a huge amount of opportunity for value adding. But what about other produce? And is value adding worth the hassle? The answer is yes...and no.

The benefits of selling only raw materials

- It gives you the ability to concentrate solely on farming and not on packaging, manufacturing or marketing
- It also gives you the ability to develop deep relationships with just one or a few key buyers.

The negatives of selling only raw materials

- Your business may end up being reliant on just a small number of buyers.
- You might need to wait a long time for payment.
- You only get the bare minimum for your product and can be screwed down on price. That is, you become a price taker rather than a price maker.
- There is nothing unique in a raw product and there can be lots if competition.
- You might only have a short window to make sales when your produce is in season.
- In the event of a major crop failure, you're unable to gain any value from damaged produce, as you could if you were able to value add it.

The benefits of selling a mix of raw and value added produce

- It gives you the ability to build a brand and diversify your business.
- You can get more for your produce.
- It connects you with the end user of your product, which is valuable for relationship building and hearing about what customers want.
- You have different ways of getting your produce to market.
- You're not wholly reliant on one buyer.
- You might be able to sell the value added products outside of the harvest season.
- In the case of hail or insect damage, you might still be able to earn an income for your crop through the value adding process.

The negatives of selling a mix of raw and value added produce

- You can lose focus on the production side of the business.
- You can end up spending too much time with customers and on administration and compliance.
- The costs to set up for value adding, either in-house or through outsourcing, can be an initial burden.

If you're interested in adding value to your raw produce, here are some of the ways you might be able to do it.

Packaging

Value adding can be as simple as putting mince in a labelled cryovac pack, pouring honey into a jar, or washing baby spinach and sealing it in a bag.

The package could be a bottle, bag, case, carton or other container. They can be standard packages or custom-made to help your produce stand out.

Changing the shape

Who would have thought changing the shape of your produce could result in a much higher return for the same fruit and veg?

Cutting, slicing, dicing and shaving your produce all classify as value adding. A look along supermarket shelves shows carrot batons, topped and tailed beans, watermelon halves and grated cabbage. Cutting and mixing your produce can also lead to value-added produce such as root vegetable soup packs.

Garlic braids are simply garlic bulbs plaited together, but they earn you a higher price for your labour than just selling the bulbs on their own. Miniature herbs and flowers are in demand by restaurants and foodies, who are prepared to pay more for them than their full-sized siblings.

Drying and dehydrating

Drying and dehydrating produce enables farmers to extend the selling season and charge a higher price for produce. The mouth waters just thinking of tucking into dried mango cheeks, dried apples with cinnamon and dried persimmons! And what about winter stews that are ever the richer for adding in dried mushrooms and beans. And for the carnivores, jerky and smoked meats have become must-haves in the pantry.

The whole herbal tea category is reliant on drying too, as are many spices and salt rubs. You can also dry produce for products such as potpourri and natural moth deterrents.

Farm dryers can range from small operations in a shed using natural ventilation through to solar and large electric dryers. Freeze-drying machines are also available.

Sprouting

Although bean sprouts and snow pea shoots have been a staple in super-markets for a long time, there's been a resurgence in people sprouting at home. This growing awareness means there's an opportunity for farmers to grow and sell unusual sprouts, as well as the seeds and kits people need to do it themselves.

Sprouting does have some very specific food safety requirements, so it's not for everyone, but trays of sprouting wheatgrass can be sold for a whole lot more than the seeds from which they emerged, and you don't need to wait for the crop to mature, so you can produce more crops per season.

Smoking

Smoking of food was historically done to preserve meats for the lean winter months, but now it's done as a gourmet service for foodies who crave the taste.

Smoked cheeses, chillies, sausages, meats, garlic, macadamias and trout are just some of the produce being value added through smoke.

Smoking set-ups can range from DIY constructions to electric cabinet devices and elaborate wood-smoke systems, all of which need to meet current food safety laws.

Roasting

Roasted nuts and seeds are a tempting snack. Roasting produce such as tomatoes and capsicums and preserving them in oil is another great way to add value to a harvest.

Crafting

If you have a crafty person on the farm, there are endless opportunities to value add produce.

Fibres can be knitted into garments ranging from baby booties to scarves and jumpers. Felted products are increasingly being seen in fashions, and fibres can also be crafted into cuddly toys.

But fibres aren't the only basis of crafts on a farm. Timbers can be carved or turned into rustic furniture. Dreamcatchers can be made from twigs and feathers. Greeting cards can be made using photos taken on the farm and herbs can be dried and bundled to create scented potpourri, garlands and wreaths.

Bottling

I remember making our first mulberry jam on the farm. Hours were spent picking the plump black-purple fruits, and then plucking out the stems. Lemons were squeezed and sugar shovelled in. The giant saucepan was like a cauldron bubbling away and blackcurrant-like stains speckled the benchtop. The smell was divine and, following the recipe to the letter, we even ran the glass jars through the dishwasher so they were clean and warm. The jam was syrupy and tangy and so, so delicious, but after half a morning's work, we only came away with six jars. The return was even less when we tried making chilli jam. Over the years though, the ability to bottle (either in our own kitchen or outsourced to others) and value add our produce helped us attract new customers and kept the cash coming in. You can value add by bottling the following:

- sauces: barbecue to chilli
- pesto: basil to kale
- jams: strawberry to feijoa
- mustards: honey Dijon to orange-infused
- pickles: crunchy to spicy
- relishes: horseradish to beetroot
- cordials: raspberry to ginger
- dressings: tangy lemon to sweet pomegranate
- preserves: lemons to marmalade
- vinegars: fig balsamic to tangerine
- oils: virgin olive oil to oils infused with garlic or lemon, and culinary oils
- pretty much whatever you grow can be bottled and preserved in some way.

FARMER SPOTLIGHT: Organic feijoas

This couple pioneered a new crop in Australia and continue to break new ground with a variety of products and ventures on their small acreage.

Farmers: Sally Hookey and Peter Heineger
Farming: Feijoas
Farm name: Hinterland Feijoas
Farm size: 4 ha (9.88 acres)
Average annual rainfall: 1 000 mm
Where: Belli Park, Sunshine Coast, Qld
Farming since: 2007

Jams, chutney, balsamic glazes, ice cream, preserved fruit, café food, pulp, a farmgate shop and a food truck have come into existence on the Sunshine Coast thanks to Sally Hookey and Peter Heineger's decision to farm feijoas.

According to Sally, Aussies pronounce the fruit as "FAY-Jo-ahs", while Kiwis say "FEE-Jo-ahs", but it's really a sexy Latin word that should sound something like "f-JSHO-as", spoken with a husky voice! So, what are they? They're a fragrant, smooth-skinned, green, sub-tropical tree fruit with creamy sweet-jelly centres. Originally from South America, they have a complex flavour reminiscent of strawber-ries, pineapple and guavas, with undertones of quince, lemon and mint.

"That's the reason they're so versatile and addictive," said Sally. "But the season is short and sharp, and they can't be stored, so we need to either sell them fresh quickly or preserve them in some way."

They sell them from land they chose for its water, views, aspect, sandy loam soil and proximity to large populations and tourist hubs. They planted 750 trees made up of six varieties, and as no one in Australia had tried to grow the crop organically on a commercial scale before, they had to learn through trial, error and travel.

"We still monitor everything and write detailed diaries daily so we can see what's working and what's not," said Sally. "We also spend time planning ahead, thinking about what direction the business needs to go in, and acting on it."

Sally's work in the newspaper industry and Peter's work as a printer brought a unique angle to their farming venture in that it's given them the ability to tell their story through the media as well as connecting directly with customers.

Working 35–40+ hours per week (80% on physical labour and 20% on the business), the couple weed, prune, fertilise, water, feed animals, do the accounts and tax, manage marketing and web sales, cook, manage stock for their farmgate shop, manage social media, plan menus and manage tour bookings. They sell their fresh and value added feijoas to chefs, at the farm gate and online.

"We do make a living from the farm, but not as high an income as an off-farm job would achieve," said Sally. "But it's great to be a part of this groundswell of change in the food-production arena. There are many, many young people interested in joining the industry for the first time in ages, and it's exciting to be a part of it!"

Food for thought from Sally and Peter

What are the basics required for your farming activity?
An understanding of soil health, basic accounting and marketing. Sally did a Nutri-tech Solutions certificate course and was so impressed she's also worked for the company. On the tourism side, since adding Myrtle the food van, hospitality and barista qualifications are also a must. We'd also recommend you join as many local networking organisations as you can.

What are the threats to the industry?
Large-scale competition flooding the market and bringing down the price. In our region, climate change could make a sub-tropical crop borderline. For small organic farmers, it's the healthiest job on earth, but it's difficult to make a full-time income and it's hard to juggle getting time off.

Opportunities in the industry?
Farmgate tourism is the hottest trend we can see right now. People are really wanting to see farm life and how real food is grown. It's great to meet people and great to grow food that gives so much enjoyment and health to others.

What makes a good farmer?
Someone who cares for the environment in their practices and has posi-

tive persistence. You also need to be adaptable, resilient, innovative and hardworking.

Best piece of advice from another farmer?
Don't rely on one market.

Worst piece of advice from another farmer?
None really – growing a new crop, we didn't get much advice!

What would you farm if you weren't involved in your current activity?
Flowers – they are so happy!

Sally and Peter's top 5 tips
1. Plan cash flow.
2. Develop your own brand and market.
3. Diversify as much as possible.
4. Take it slow and enjoy the ride.
5. Adapt to current trends.

What's the best way to enjoy your product?
Feijoas are super versatile, but my favourite recipe is feijoa ice cream!

Fermenting

Fermented foods are those that are produced or preserved by microorganisms. Zymology is all about the science of fermentation, and has been harnessed by many a farmer to add value to and dress up their produce. What was once a traditional way of making foods last and increasing their health benefits has been rediscovered and embraced by Australian farmers and consumers.

Hot items at farmers' markets include artisanal sourdough breads, traditional sauerkraut, Korean kimchi, kombucha drinks and kefir. Many cultures from around the world have their own fermented food traditions, and these can be explored to help you bring a specialty product to market.

Growing papaya? Pick them unripe and grate them for pickles. Too many mangoes on your hands? Try the Filipino recipe for Burong mangga. Growing soybeans or rice? There are many fermented foods (and beverages!) to explore.

Fermenting foods is rich in tradition, and as a way of adding value to your produce, potentially rich in reward.

Flowers

Although some people never buy flowers for their home or gifts, those who do are prepared to spend regularly. Flowers are seen as a traditional and 'safe' gift and can brighten up a home or brighten a friend's day. They are also being used more frequently in real estate industry sales campaigns with houses festooned with them on open days. For these reasons, flowers can walk out the door at farmers' markets. They can also be sold direct to florists and the big central markets, as well as at the roadside. They can be grown for special events such as weddings, be dried and added to teas and are also used in culinary dishes. From tiny bouquets of lavender to giant sunflowers, from bouquets of Australian natives through to festive wreaths, flowers can be value added in numerous ways from just tying a ribbon to gathering different colours together and presenting them well.

The Australian flower industry is large, and there are numerous ways to value add using focal blooms and foliage.

Animal hides and horns

The hides of cattle can be value added into floor and furniture coverings. Sheepskins can warm the feet as Ugg boots, and are often used in baby's rooms, while crocodile skins are feted around the world as a luxury form of leather.

Animal horns can also be used not only as a receptacle for biodynamic preparations, but also as pieces of art, for musical instruments or, in the case of deer antlers, for hat racks. If your beef breed has horns, you could even start up your own premium line of buttons!

Alcohol

Grapes make wine, but have you ever thought of value adding other crops into alcohol? The humble potato can be turned into vodka, elderberry can make a delightful light champagne, and rice is the basis of sake, but there are so many other options open to farmers.

Jaboticaba is a Brazilian tree, distinct in that the marble-sized purple fruit grow directly on the trunk rather than from the branches. It can be eaten fresh, made into jam, or turned into a great liqueur.

Beers and wines can be flavoured with chilli, gins can be infused with Australian natives and hops are needed in ale making. Honey is the basis of mead, and there are endless plants around the world you can research to see how they have been fermented into alcohol. It's likely that if you can grow it, you can use it in some way to make alcohol. We once made 60 litres of beetroot wine that tasted like a velvety red. However, the next 60 litre batch tasted like a mix of dishwater and bleach!

When it comes to agricultural alcohol, you're only limited by your imagination, science and liquor licensing laws.

There are other opportunities if you are producing alcohol in that you can provide a customised label service for corporates, weddings and special events. By offering a personalised labelling service, you can charge a premium just for changing the sticker on the front of the bottle!

Beauty and household products

Opportunities abound in the beauty and household product space for farm-produced products.

Thinking about orange farming? How about value adding what many see as a waste product by producing a natural orange peel kitchen cleaner? Have your own bees? Beeswax can be the basis of your balm line or a key ingredient in your furniture polish. Goat milk? There's obviously soap, but face masks can also be made fresh. Herbs can be dried and mixed with bath salts, and you can make natural air fresheners, insect repellents and so many other items from farm-grown produce.

Specialty diets

Value adding foods by targeting specific groups can be a good way of differentiating your produce in the marketplace and reaching new customers. You could investigate value adding in the following areas.

- Organic
- Biodynamic
- Gluten free
- Diabetic
- Vegetarian
- Vegan
- Paleo
- FODMAP (a diet for people with food-intolerance issues)
- Common allergens (such as nuts, eggs, dairy, wheat, soy)
- Asian foods
- Super foods
- Kosher
- Halal
- Healthy pet food

Your value adding might be as simple as a well-designed label targeting a key group, or perhaps you will blend certain products together to make something unique.

Information and intellectual property

Information is another way to add value and increase sales of your produce.

When our turmeric was ready to harvest, we created a simple photocopied booklet with tips on how to plant, grow, store, cook and use it for health purposes. The booklet helped us to sell more turmeric for the following reasons.

- Displaying the booklet in a prominent location enticed customers to come closer to the stall and engage with us.
- The information contained in the booklet showed people all

the different ways to use and store it, so rather than just
buying one rhizome, they were happy to buy 500 g or more at
a time.

- Providing the booklet also built customer loyalty, with many
  returning to buy more turmeric from us season after season.
  It also saved us from having to explain everything we knew
  about turmeric to every customer, which freed us up to sell
  rather than just inform.
- Finally, we also made money by selling the booklet for double
  what it cost us to print.

Think about the knowledge you have (or will have), and how you might
be able to add value to your produce with it. Some potential opportunities might include the following.

- Recipe cards for in season produce
- A cookbook
- How-to booklets on using your produce in unconventional
  ways
- An audio or printed book about your farming journey
- A DVD, podcast or YouTube channel
- Doing (or having a chef do) a cooking demonstration at a
  market
- Running a course, either in person or online
- Pictures from your farm turned into postcards or sold via
  photo sites

Value adding is something you, as a small producer, can undertake to
extend the harvest, entice new customers, make more money and make
the most out of a crop. Think creatively, and entire new revenue streams
and opportunities could open up for your farm.

～

FARMER SPOTLIGHT: Chestnuts and pigs

A couple in their 60s wanted to produce food, reconnect with nature, and become part of the solution to what they saw as increasing foreign ownership of Australia's prime agricultural land.

Farmers: Linda and John Stanley
Farming: Sweet chestnuts, chestnut-fed pork
Farm names: Chestnut Brae
Farm size: 29 ha (72 acres)
Average annual rainfall: 944 mm
Where: Carlotta, Nannup, Western Australia
Farming since: 2013

Linda and John Stanley spent two years looking at farms before selling their home in Perth and signing the title for Chestnut Brae. The Stanleys needed to continue their retail consultancy work to support the farm's development, so one of their criteria when looking at land was that it needed to be within a three hour drive of the international airport so they could continue to service their overseas client base. They were also looking for a farm with good, undulating, arable land for future vegetable production, permanent water, good mobile phone reception and – being older entrants to farming – the opportunity to be able to add value to an already mature crop.

"We wanted to create unique, niche products rather than being a 'me-too' farm producing 'me-too' products," said John. "That's where the sweet chestnuts came in, as no one else was growing them in Western Australia and they can be value added into everything from chutneys to sauces. They were already growing on mature trees, so we didn't have to wait to start production."

There was a lack of knowledge in Western Australia about chestnuts, so the Stanleys did their own research, travelling to chestnut-growing regions in Italy, France and Corsica. They returned with lots of ideas and 300 recipes. They were stunned to discover that the bulk of Australia's harvest from the east coast is sent to China for peeling, packing and freezing before being imported back into the country. Committed to local food production and processing, they invested in their own peeling machine and now make chestnut conserve, chestnut puree, chestnut red onion and fennel chutney and Bloody Mary Chestnut Ketchup and vacuum-pack raw and roasted chestnuts on farm. They also sell chestnut-fed pork, and have recently converted the old house into a farmstay.

The Stanleys' typical 60+ hour week (30% physical labour and 70% admin) involves feeding and watering animals daily, mowing grass and orchards, shifting sheep and goats in their electric fences, desuckering chestnut trees and cooking, jarring and labelling chestnut products. There's also the weekly farmers' market to attend and general tasks such as updating social media, corresponding with chefs and doing the books.

"The physical work is exhausting, but it also burns off the mental stresses of our consultancy business," said John.

"Don't enter farming if you think it is an easy get-rich-quick opportunity," said Linda. "Enter it because you have a passion for something related to farming or producing good food and then follow your heart. We do miss the regular company of our family and grandchildren, who are three hours away, but building the business is a fun challenge, and the chestnut orchard is like a European forest – it's magical."

Food for thought from Linda and John

What are the basics required for your farming activity?
Research skills, so you can find out all about nut production and potential produce. It can also be a great help to join a local group interested in the way you want to farm. For us, that's regenerative agriculture techniques and organics. We've found that networking with other farmers also helps us to market and sell products.

What are the threats to the industry?
Government overregulation of farmgate sales can be a challenge, and economic downturns make everyone tighten their purse strings, which for some means that small-farm-grown organic produce is off the menu.

Opportunities in the industry?
The rising interest in local food, organic food and food purchased direct from farms is a big opportunity. We're capitalising on that by working to

build a regional food trail and spreading the word about organic farming.

What makes a good farmer?
Someone who's steadfast and keeps going no matter what. That said, you also need to be flexible and try doing things differently if it doesn't work out the first time. There's no getting around the fact that farming is hard work, both mentally and physically. A good farmer appreciates that one of the rewards for all that hard work is a connection with nature.

Best pieces of advice from another farmer?
Use pig fencing (Waratah, not the cheap products that don't last) to deter kangaroos. And use a horse float rather than a trailer to load pigs to take to slaughter.

Worst piece of advice from another farmer?
"Do nothing in the first year and see how the farm goes." We didn't take that piece of advice.

What would you farm if you weren't involved in your current activity?
Chestnuts lower cholesterol, are low GI, low fat and taste scrumptious, but if we weren't farming them we'd probably farm avocados so we could obtain avocado oil and make skincare products.

Linda and John's top 5 tips
1. Fencing is a huge cost – check the fences and make sure that they are in good shape. Nobody told us this, and our fences were in a terrible state, but we didn't see it. It cost us a lot of money to bring them up to standard.
2. Don't be traditional – be unconventional. Look at what Joel Salatin in the USA is doing on his Polyface farm, and soak up inspiration from the Salatin story.
3. A tractor is a huge help, so if there isn't a reliable tractor on the farm you plan to buy, consider getting one as soon as possible.
4. Read widely about the crops or products you are thinking of implementing first to see what is involved.

5. Network with other like-minded farmers to get your produce to market and promote your farm.

What's the best way to enjoy your product?
Freshly roasted chestnuts with a glass of red wine, or try John's favourite, Chestnut honey ice-cream.

Ingredients
1 teaspoon vanilla extract
2 cups of heavy cream
1 cup milk
3/4 cup raw honey
1 cup chestnut purée – peeled and steamed chestnuts pureed with ½ cup milk
4 large egg yolks

Preparation
You'll need an ice cream maker or a Thermomix.
Add vanilla, cream, honey, milk and chestnut purée to Thermomix and process until smooth.
Add egg yolks to cream mixture in a slow, steady stream, processing constantly in the Thermomix.
Pour into ice-cream maker and process until soft-frozen, 20 to 25 minutes, then transfer to an airtight container and put in freezer for about three hours to harden.

PART III

10

THE GETTING OF WISDOM

If you were born into a farming family, you've been absorbing knowledge since birth, but for the rest of us, it's time to catch up. When I started farming, I didn't know fireweed from turmeric, nor how to milk goats, care for bees, trim alpaca hooves, or identify Chinese jujubes, let alone grow, dry and sell the fruit, but I wanted to learn, and it ended up being easy to find out.

To speed up your knowledge accumulation and practical skills, there are numerous pathways you can take. Here are some of them.

Books, the internet and the media

Want to cultivate mushrooms? There's a book for that. Want to see a hundred different ways to catch a bee swarm? There's YouTube for that. Want to know how to grow organically? There's a magazine for that.

Whatever your interest – whether it's farmhouse cheeses, growing berries, the natural care of goats or how to build a compost heap – there have never been more resources available for the beginner farmer. That doesn't mean you can trust everything you read on the internet, or that it will be appropriate for your region's climate, conditions, laws or marketplace. So, read widely, compare information, do your own thinking and trial any advice before you fully commit to it. There are some useful Facebook groups to join too.

Sometimes, information in books can be even more focussed and in-depth than the free information available on the internet. Books specific to your needs will be a truly amazing resource you will return to time and time again. Our own bookshelves are stacked high with titles from US publisher Chelsea Green, as well as plenty of self-published titles in various areas of interest. You can buy books, borrow them from libraries and swap and share with neighbours too.

ABC's remarkable *Landline* program is a must-watch for anyone interested in farming. ABC Radio also offers fascinating rural reporting, and you can learn everything from the current market price of pome-granates to available rural grants from newspapers such as *The Land*.

Associations and industry organisations

From amateur beekeeping associations through to professional associa-tions for breeders of miniature pigs and buffalo, there's sure to be a group of people interested in the same farming activities as you. Associa-tions provide opportunities to meet other like-minded people, source and sell breeding animals, participate in combined marketing opportu-nities and learn valuable information from association publications and events. See www.smallfarmsuccess.com.au for comprehensive links.

Field days

Field days such as those at Tocal and Henty in NSW gather together breeders, machinery firms, fencing suppliers and experts you can watch, listen to and talk to about everything from specific sheep breeds to how to control weeds. Government organisations and private companies attend these field days armed with plenty of information to share and sell.

There are also field days literally spent in fields learning from other farmers how they manage their land, soil, biodiversity, fencing, water-ways and livestock.

Conferences

If you want the latest information on your industry, attending a conference can be a great investment. Conferences give you access to the latest scientific information, and there are normally sessions on everything from marketing opportunities to industry threats and future thinking. At conferences, you get access to the best and brightest sellers, buyers and middlemen, and you can make contacts that will assist you in building your networks and your knowledge.

Conferences are held both around Australia and internationally, and the good news is that if you're farming commercially, the cost of attending them is tax deductible. From tropical agriculture to agri-tourism, from beverage crops to dairy science, you'll be able to soak up a world of information quickly.

Farmstays

What better way to undertake research than by combining it with a holiday in Australia or overseas! From camping to luxury accommodation, there are opportunities to stay on farms running huge sheep and cattle operations down to smaller farms growing apples and producing cheese. For a small investment, you'll get a taste of country living and spend time with a farmer doing what you want to do.

Agricultural tours

Agricultural tours offer a more formal approach to a holiday in that you explore a region, but with a focus on learning about a specific industry, be it aquaculture, rice-growing, or beef or dairy operations. You can tour within Australia or join overseas study groups if you'd like to see what's happening around the globe.

Landcare

We joined Landcare when we first moved to our farm, and it was one of the best things we ever did. In the early years, it was a huge help to be able to access the knowledge of other members. We made friends, successfully applied for grants to protect riparian (creek) zones on the farm by fencing out livestock and planted a one kilometre wildlife corridor and windbreak of native trees and shrubs along the south-western boundary of the farm. Our Landcare group held meetings and information sessions on everything from weeds to compost-tea making. One of the big bonuses was indulging in all the home cooked goodies that the members brought to each meeting!

With an estimated 6 000 groups involved in Landcare, there's sure to be one operating near the area you intend to farm, or where you are farming. Through Landcare, you can also access grants and information from other agriculture and water authorities. It's a volunteer organisation, so there can be lots of different personalities and opinions involved and groups can have an organic or chemical-farming focus. It's worth popping along to a meeting to see if you can tap into some inspiration and advice that will help you in your venture.

Seedsavers

If you are interested in plants and in getting to know other like-minded people, Seedsavers is a wonderful group that will help you meet people quickly while also giving you access to traditional, open-pollinated (non-hybrid) seeds that will do well in your area. Seedsavers get-togethers are often held at different members' properties, which enables you to see other gardens, farms and ways of producing. There is often an exchange held at each meeting where people swap plants, but if you're not up and running with your plants yet, you can always bring something home-cooked or homemade. And if you find that there isn't a Seedsavers group in your area, you can always set one up!

WWOOF/Helpx/Workaway

Nothing beats hands-on knowledge, so getting work experience before you start spending money on your venture is a worthwhile investment. By working alongside another farmer, you exchange your help and time for training, food and board.

The three main schemes available are World Wide Opportunities on Organic Farms (WWOOF), HelpX and Workaway. There are thousands of hosts throughout Australia and around the world, so you can choose a farm based on either interest, location or both.

Do your research well before contacting a host (including reading between the lines of past volunteer reviews), and be up front that you are volunteering because you want to learn about the industry. Each host will have different requirements regarding the length of time they require you to stay, the duties you will perform and the type of accommodation and food available. Stays are generally for periods of between one week and three months.

While you are volunteering, be respectful of the fact that you have been invited into someone's home and business. If you want to suck a farmer dry of information, make sure you are also sucking yourself dry helping them every day! Be considerate, too. If you're planning on setting up a banana farm at Mackay, perhaps look to doing your work experience with a host in Coffs Harbour so that the Mackay farmer doesn't feel uneasy providing information to a potential competitor. Then, perhaps volunteer on a different type of farm in Mackay so you can tap into the regional information you need.

Internships

Some small farms offer internships as a way of dealing with labour costs and in the spirit of helping newcomers to the industry get the training and experience they need. Internships often require a longer time commitment from you, somewhere in the order of three to six months, sometimes even up to a year, but it will enable you to learn an operation across one or more seasons. When seeking an internship, ensure that there will be proper training and diverse learning opportunities so that you're not just left to the weeding – although this can be a big part of market garden type operations.

Private courses

There are private trainers operating in all states, both in the city and the country, offering a variety of workshops and courses. Permaculture, beekeeping, mushroom cultivation, cheesemaking, soap-making, fish farming, pig farming and other skills can be learnt this way for a price.

There are also courses in holistic management and programs such as GrazingforProfit that take a holistic business, earth and people approach to farming including empowering you to define your goals, take control of your enterprise, develop a drought management plan, increase rainfall use efficiency and strengthen family communications.

Government training solutions

Each state government offers some form of specific rural industry training, whether through the provision of extensive research notes and case studies or through workshops and courses. See this book's website for links.

In New South Wales, the government's Department of Primary Industries offers courses in regional areas under the ProFarm brand. These courses are specific, are often delivered over periods ranging from several hours to a few days and offer learning in very specific areas ranging from alpaca care and management to pig nutrition, chainsaw use to queen bee breeding, irrigation to introduction to pastures. They also offer courses in business structures for rural family businesses, including succession planning.

TAFE and other colleges

The TAFE system provides learning opportunities in a diverse array of agriculture subjects. This can lead to industry-recognised certification that may assist you in gaining employment off your farm, as well as better equipping you to run your farm. TAFE offers some courses online and many others face to face. TAFE courses are normally shorter in duration and less expensive than university degrees, but longer than private and government backed short courses. You can learn about and be certified in areas such as but not limited, to those listed below.

General farm duties: Administering medication to livestock; chemical use and application; assistance in artificial insemination of livestock; care and training of working dogs; fencing; animal care; operation of farm machinery (front-end loaders to chainsaws, quad bikes to tractors); installing, laying and maintaining irrigation systems; mustering and moving livestock; weed control; welding; tree felling; using firearms to humanely destroy animals.

Sheep: penning; shearing; wool-clip preparation; wool handling; wool classing.

Horses: farriering; handling young horses; educating, riding and caring for horses and equipment.

Dairying: carrying out birthing duties; castration; collecting, storing and administering colostrum; operating effluent-recycling systems.

Horticulture: transplanting small trees; seed collection and propagation; general crop establishment, maintenance and harvesting; potting up plants; planting; pruning; treating diseases and pests; soil testing; arboriculture.

Aquaculture: hydroponic systems; producing algal and live feed; handling stock; food and hygiene practices; hatchery operations; environmental monitoring; navigation; operating vessels; slaughtering livestock; net maintenance; servicing of propulsion units.

Beekeeping: using a smoker; constructing and repairing hives; managing swarms; requeening a hive.

There is also the NSW DPI's Tocal College which offers training and pathways for school leavers through a residential college. It also provides delivery of PROfarm short courses and educational texts.

Universities

Universities across Australia offer degrees in agriculture and are open to mature-age students as well as school leavers. These three year or longer courses offer a mix of academic study and practical and skills. To give you an idea of what's on offer, Charles Sturt University has degree courses in areas such as sustainable agriculture, viticulture and oenology, wine business, animal science and equine science.

Other ways to learn

You could invite locals over for dinner to share stories, food and knowledge, or pay a consultant to answer your questions. You could offer to reimburse a farmer for time spent learning from him or her or volunteer to work on their farmers' market stall. You could also join the Country Women's Association (CWA).

Trial and error is an impactful learning method, but best avoided if possible by asking questions, actively listening and learning from the successes and failures of others. With all this help at hand, there's no need to be overwhelmed by your lack of knowledge of agriculture. Besides, there are immense benefits in being fresh to the industry in that you are not stuck in the antiquated groove of "my father did it this way, his grandfather did it the same way, and it was good enough for his father too." By going out and seeking information, voraciously absorbing it, and then putting it to the test, you'll be well on your way to success.

FARMER SPOTLIGHT: Biodynamic mixed vegetables

A husband and wife team, inspired by international chefs, farmers and activists, plant their hands and hopes deeply in the Australian earth.

Farmers: Falani and Olivier Sofo
Farming: 50+ species of mixed vegetables
Farm name: Living Earth Farm
Farm size: 2 600 m of cultivated space on one acre of leased land on a larger biodynamic farm
Average annual rainfall: 800 mm
Where: High Range, Southern Highlands NSW
Farming since: 2016 on this land, and 10 years around the world prior to that

Olivier's Italian parents brought their green fingers with them to Australia, and from an early age he helped tend the garden and make preserves. Apprenticing as a chef at the age of 15, he was swept up by the restaurant world, moving to Paris to work under French chef Alain Ducasse. Fifteen years in the industry followed, but by 2010 Olivier was ready for something different.

"I moved back to Italy and became inspired by an incredible biodynamic vigneron, Stefan Bellotti," said Olivier. "He exemplified what biodynamics and natural systems farming looked like after 40 years. You asked him about wine and he spoke about plant and human health. He inspired me to farm in this way."

Olivier then interned with organic market gardeners Michael Plane and Joyce Wilkie at Allsun Farm in Gundaroo, NSW, before heading overseas to Michael Ableman's Foxglove Farm in Canada. This is where he met his match in Falani.

Falani, a Canadian, also came to farming through her work as a chef, inspired by the taste of fresh produce. She apprenticed with chef, farmer and activist Michael Stadtlander at Eigensinn Farm in Canada, worked at a sustainable dairy in Wales, and spent two years at Foxglove Farm. There, it wasn't just the idea of sustainable agriculture that was growing in her, but also her love for Olivier and their desire to establish a farming enterprise of their own.

The couple acted on Michael Ableman's call to action, 'Agriculture is facing a crisis of participation.' They decided to dive even further in, moving to the Southern Highlands in NSW, where they put a 'looking for land' advertisement on their local Small Farms Network website. The ad led to discussions with a young family who were farming biodynamic beef, and their matching philosophies saw them offered a lease of one acre of land with deep, clay loam soil, good aspect and water access.

"We have a very positive relationship with our land partners and there's room to expand," said Falani, something they might need to do given the demand for their bountiful harvests of everything from Chioggia beetroots and Delicata squash to French melons and radicchio.

"This is an industry where you must wear many hats, those of a builder, a grower, a biologist, a marketing specialist, a salesperson, a delivery person and an artist, all with enthusiasm and as much confidence as you can muster and drive," said Falani.

With baby Naia now on the scene, the Sofos work for 40–50 hours per week between the two of them (85% on physical labour and 15% on the business). Olivier also works off-farm for a day or so a week. A typical week involves two days spent cultivating, planting, preparing beds and undertaking general maintenance. Other days are focussed on harvesting, organising and delivering orders, updating the online shop and endless irrigation tasks. They sell their produce online to local customers and grow specialty crops for chefs.

"The work is so grounding and satisfying," said Falani. "We breathe fresh air, we move our bodies, we eat well. We have moments of silence for reflection. We get to raise our children together outside. We get to work together and be creative and solve problems. We also feed our community, and watch the process of birth, life and death over and over each season."

"There are not many jobs out there where you get up every day and go to work with the intention to give more than you will ever take. A good living does not just refer to having a good net profit or hourly rate. It also means that you are doing 'good'."

Food for thought from Falani and Olivier

What are the basics required for your farming activity?
Passion, diligence and a 'give it a go' attitude. Be willing to ask questions, observe what's happening in front of you and read books such as Elliot Coleman's *The New Organic Grower*, Pam Dawling's Sustainable *Market Farming: Intensive Vegetable Production on a Few Acres*, Ben Hartman's *The Lean Farmer*, Allan Savory's *Holistic Management Handbook*, Pat Coleby's books, Alex Podolinsky's *Biodynamic Manual*, and books by Wendell Berry and Rudolf Steiner. We'd also recommend presence, intention, BD 500 and a regular foliar application of seaweed, humic acid and fish hydrolyslate.

What are the threats to the industry?
Difficulty accessing high quality seeds due to corporate seed companies, the size of the industry in Australia and import restrictions.

Opportunities in the industry?
Natural farming is on an upward trajectory. There is so much room to

move and grow.

What makes a good farmer?

A good farmer sees the big picture and is respectful of how farming works within the ecology of the landscape. Small, diverse farmers also need to be entrepreneurial, creative, dynamic and open to change. Having a clear vision and solid philosophies help when you experience failures, and make you resilient enough to keep at it. A good farmer is also a healthy one, so take care of your body or else it stops. Farming can take its physical toll.

Best piece of advice from another farmer?

As a small farm, the way to make it work financially is to glean the fields and sell absolutely everything.

Worst piece of advice from another farmer?

Just spray the whole orchard, don't bother with hand weeding.

What would you farm if you weren't involved in your current activity?

Trees, cattle and pigs. There's nothing better than planting a tree, cattle are marvellous animals and pigs are tenacious and beautiful.

Olivier and Falani's top 5 tips

1. Work on farms and with farmers who inspire you.
2. Observe what is happening in the fields and take notes.
3. Spend time looking for the right land base to suit the farming you wish to do.
4. Understand who your customers are, or might be, before starting production and value those relationships.
5. Be courageous, and don't be afraid to try new things.

What's the best way to enjoy your product?

We love preparing our food simply and adding a raw component to our meals such as shaving vegetables and dressing them with olive oil and vinegar, or if it is a sweet Ailsa Craig onion, just caramelising them and serving on top of roasted potatoes.

11

PLANNING AND GOAL SETTING

hat do you really want out of this whole farming venture?
What is it that you want to achieve? What is your overall
vision? How will you bring it to life?

If your reason for farming is to relax and get closer to nature, you'd
be crazy to set up an intensive pig-farming operation. If you want to
make enough money to put your children through university, you'd be
just as crazy to try and farm a couple of heritage pigs.

For your dreams to align and become reality, you need to plan and
set some goals. Once you've determined these goals, commit them to
paper. This will help you stay on track and, importantly, alert you to
when you've achieved them.

There have been a gazillion great books written about business planning and goal setting, so here are just a few starting points to think
about.

Financial goals

- How much do you want to earn? Why?
- How much do you need to earn in the first year and following
  years?

- How much are you prepared to plough into the business?
- What kind of off-farm income are you prepared to commit to supporting the venture?
- What research have you done to support your projected figures?
- What type of working capital are you able to arrange so you can go into farming with the liquidity you need to make it work?
- Have you planned for bad years and crop failures?

By planning your venture and setting your financial goals, you'll discover better ways of doing things. For example, let's say you've sold your house in the city and are buying a cheaper farm in the country. Does it make sense to buy the farm outright, or might it be better to keep a mortgage on the farm so you have access to lower-interest-rate loans compared with business loans and credit cards that can cost upwards of 12%? Speak with a professional so you can also get insights into how not to overcapitalise, e.g., rather than building more infrastructure on your piece of land, it might make sense to build it in an industrial area or on another farm so you are spreading your assets across other properties. You might even be advised to just rent it.

Finally, what is your exit plan?

Business goals

What kind of business will you be proud of running? What do you want to get right that you have seen so many businesses get wrong? The beauty of goal-setting for your own business is that you are in the box seat to create the company you've always envisioned. For example, do you want to be known for the quality of your produce? If so, a goal like this means you need to focus on healthy soil and quality control. If you want to be known as a good workplace, this will mean extra thought on human resources, hiring and workplace issues. If you want to be known as an innovator, then plan to stay up to date with information from around the world, feed your creativity and take risks.

Experiential goals

What goals do you have when it comes to experiences? What values, skills and knowledge do you want to gain? What's the ride you want to go on?

Think about what you want to learn, feel and experience during your farming adventure. It might be something as simple as making time to watch the sunset over your land three days a week. It might be that you want to attend an overseas farming conference to absorb information from different cultures. It might be that you want to experience what it's like to be self-sufficient or part of a vibrant community, or to kill and prepare your own livestock.

Physical and mental health goals

What goals do you have for your physical and mental health?

Perhaps you want to trim down and eat more simply. Perhaps you want to focus on living with joy and not burning out. Perhaps you want to stay physically strong into your seventies. You'll only achieve these things if you work these goals into your day-to-day life and workplace health and safety procedures.

It's important to think about your physical and mental health goals, integrating them into how you run your farm, because there's no use giving everything you've got to make the farm work if it leaves you run down and/or depressed. There's no use being super fit if you ride a quad bike in a careless manner and roll it, and you'll be of no use to anyone if you lift heavy items on your own and injure your back.

Relationship goals

What hopes do you have for your relationships, and how can you enshrine these hopes in how you run your farm?

- Want a better relationship with your partner? Plan for time away from the farm and the drudgery of chores. Think about holding formal business meetings rather than constantly talking work when you could be talking romance.

- Want to create a strong, enduring family? Pay attention to fairness, fun and freedom, and begin succession planning early.
- Want to be known as 'good people' by locals, suppliers and employees? What will you need to do to be an authentic, welcomed member of the community? Tip – among other things, it will include paying your bills on time.
- Want to stay in touch with your city friends? Plan for fun working-bee weekends at the farm, and getaways to other places where no one needs to do a thing other than socialise.

Environmental goals

How do you feel about the environment and how do you want the environment to feel about you and your farming and lifestyle impacts? How will you treat the land, the water, the wildlife? Do you want to pillage for profit, maintain or improve? What are your goals when it comes to biodiversity?

Legacy goals

A legacy is something that is left behind after you die. What legacy do you want to create and what is the best way to create it?

Do you want to leave fun memories, inspirational lessons or a pot of gold – or all three?

How will you have impacted your family, your community, the environment and those yet to be born?

Basically, how do you want to be remembered, and what steps do you need to take to make a positive impact while you're here and leave something special behind after you're gone?

Some people are great at planning and goal-setting, while others just like to wing it. You can do everything from a SWOT (strengths, weaknesses, opportunities, threats) analysis to setting SMART (specific, measurable, attainable, relevant and time-bound) goals. You can create a vision board (how you want your life to look in pictures) and use NLP (neuro-linguistic programming) techniques to spur you on – but don't wait until you've got all your carrots in a row – planning without action just leaves bare fields and barren dreams. Get on with it and get going!

FARMER SPOTLIGHT: Biodynamic beef, prime lambs and Saddlebacks

From biscuits to beef and biodynamics, the Arnott family have been a part of Australia's food heritage since 1865.

Farmer: Charlie Arnott
Farming: Shorthorn cattle, merino and prime lambs, Wessex Saddleback pigs
Farm name: Hanaminno
Farm size: 2 143 ha (5 000 acres)
Where: Boorowa, South West Slopes and Plains, NSW
Average annual rainfall: 625 mm
Farming since: he was born

Charlie Arnott is the great, great, great grandson of William Arnott, the founder of the Arnott's biscuit company. The family who gave us Iced Vovos, Monte Carlos, Tiny Teddies and Tim Tams was bought out by America's Campbell Soup Company in 1997, but 27 years prior to that in 1970, Charlie's father had already got his wagon wheels on, leaving the biscuit company's factory floor for the farming fields of Boorowa.

Those fields now rear Shorthorn cattle, with around 300 breeders (half the herd size of years past but with weaners grown out on the farm) as well as 1 000 fat-lamb merinos and first-cross ewes.

"The Shorthorns are versatile, reliable and good mothers with a great temperament," said Charlie. "They're grass fed with some supplementary straw cut on the farm."

Charlie grew up on the farm, went to boarding school and university, and then returned to take over the farm in 1997. These days, he has a young family, has employed a farm manager and works at the farm for 10 days a month. From a coastal base, he leads other projects involving farming, food and ecommerce including *FoodBomb*, a wholesale food marketplace for cafés and restaurants, and *Butcherman*, an online marketplace for farms, butcher shops and meat wholesalers.

"My parents had been very supportive with succession planning, and around 2004 I started really thinking about where the farm was going," said Charlie. "Farming conventionally had been about producing lots of wheat, wool and beef with lots of inputs. But I attended some courses that really turned my head on and I realised the future for us was about producing a product, not a commodity. I needed to farm with things like the environment, family, soil, staff, livestock and legacy in mind. Organics and biodynamics just made so much sense."

Charlie credits the RCS GrazingforProfit program, the learnings of people like holistic farmer Allan Savory and biodynamics educator Hamish McKay, and a farmer support group he attends as central to his growth as a farmer and businessman.

"The programs were a big pivot for the business, the physical farm and my head," he said. "Meeting regularly with other family farming businesses has been supportive and life-changing. I've learnt that farming is all about intention, and that 'lazy' farming – working with nature – is smart farming."

When Charlie's at the farm he works 60+ hours per week (80% on physical labour and 20% on the business). Charlie sells direct to butchers in Sydney and online, with 20 kg boxes of beef shipped to customers.

"The farm is my sanctuary. It's grounding, and it's the place where I'm happiest," said Charlie. "But my other passions are social and cultural enterprise. I'm trying to create a legacy for my children, the community, and the customers who eat our food. It's about creating new attitudes to farmers."

"I love asking people who their doctor is. They can always give me a name, and then I ask how important the doctor is to them. Most say, 'very important'. I ask how often they see their doctor and they say, maybe every two months. Then I ask them who their farmer is. They shake their head, and then it slowly dawns on them that they need their farmer three times a day."

Food for thought from Charlie

What are the basics required for your farming activity?
I did a Bachelor in Rural Science, but you don't necessarily need formal qualifications. What you need are skills in stock handling, a mentor and if you want to develop further, a program like RCS's GrazingforProfit. You also need to have a good understanding of soil biology.

What are the threats to the industry?
Ageing farmers, lack of succession planning, the difference in expectations of the younger generations and the rise of corporate farms and technology means there are fewer families in rural communities. Amenities reduce, and once you start picking that thread it unravels quickly. That's because when families are removed from communities, a large chunk of the community's ability to thrive disappears too. It doesn't matter how good technology is, it results in local job losses, and that's a disaster because technology can aid you but it can't replace human spirit and endeavour.

Opportunities in the industry?
Matching young, keen and curious people with the resources and assets of older farmers so we can reinvigorate rural communities and land. Succession planning doesn't just have to be about family members. It can be about bringing people into the enterprise who are intelligent and raring to go. There are also opportunities in paddock to plate, with more people wanting to know where their food comes from and who it was grown by.

What makes a good farmer?
Resilience, courage and curiosity, and the understanding that progress is better than perfection. If you're too much of a perfectionist it can be a trap, so it's moving forward knowing that 80% of something is better than 100% of nothing.

Best piece of advice from another farmer?
The best fertiliser you can put on your own country is your own footprints.

Worst piece of advice from another farmer?

An agronomist once told me that "you can't farm without chemicals," but I try not to hang around people who give bad advice. In the end, we're the product of the people we hang out with the most.

What would you farm if you weren't involved in your current activity?
I'm keen to create a multi-species farm model that could be rolled out to feed whole towns. It would involve social enterprise, livestock, pastured eggs, permaculture, food forest areas and more.

Charlie's top 5 tips
1. Ask better questions. When approaching a new situation, new enterprise or problem, always start with 'why?'
2. Don't go around in circles worrying about what other farmers are thinking about what you're doing.
3. Start with the end in mind – have a long-term vision and short-term focus. Be clear and methodical about why you're doing it.
4. Don't adopt, adapt. Things aren't prescriptive; you can't just apply a solution, you need to adapt it to your specific circumstances.
5. A problem is an unmet need. Solve problems by identifying the needs of the people or elements involved. This applies to farming and to life.

What's the best way to enjoy your product?
My wife's amazing osso bucco.

12

EMBRACING TECHNOLOGY

You might have a vision of a completely off-grid farm where the only horsepower you use comes from a team of 17-hand Percherons, but for other types of farms it pays to look at what technology is available, as it can have a big impact on how you run your farm. Although some technology and machinery can be unnecessary overkill that burdens you with unnecessary debt, the right options for your farm could see you tapping into technology to improve efficiency from preparing the soil right through to marketing.

The advent of technology has changed agriculture massively. Here are just some of the ways in which it has had an effect.

- Irrigation. You can set up an irrigation system, program it into your phone and turn it on and off from your lounge room or even from overseas (but you'll still need a neighbour or someone close by in case a pipe breaks). Solar pumps on dams don't need the regular refuelling that diesel pumps require.
- Drones. You can fly over your property rather than riding over it. Drones can be used to monitor crop health, spot irrigation leaks, monitor sunlight absorption rates, note where fertiliser is needed and check on livestock.

- GPS. No longer do you need to peg out and line up for planting and harvest, you can program a tractor with GPS and drive it in a dead straight line for 5 kilometres.
- Sensors. In-ground sensors can be used to alert you to moisture levels. They can even be implanted in cows to let you know when she's ready to mate.
- Robots. Don't want to get up early to milk the herd or pick the fruit? There's an agbot for that.
- Machinery. Whether it's a labelling machine or a milking machine, there's something to make your life easier. Tired of twisting caps on by hand? There's a lidder. Want the perfectly measured squirt of jam? There's a piston-filler. Sick of hand weeding? Use flame or steam machines.
- Weather apps. These alert you to developing conditions well before they affect your farm, so you can plan ahead.
- Price research. With the tap of a few keys, you can find the latest pricing at the major markets for your products.
- Fencing. Portable, electric fencing used with solar panels enables economical paddock rotation.
- Security. Camera systems and motion detectors can alert you to everything from wild dogs to unwanted human visitors.
- Social media. This enables you to connect with customers you didn't even know wanted your product.

If you are interested in using machines and technology on your farm, pay close to attention to maintenance and lifetime ownership costs. Additionally, do you really need to own all the equipment yourself if it is just going to lie idle for much of the year? Is there a way you can share it with neighbours or rent it out when you're not using it? The Stayz and Uber and Airbnb of agriculture...yep, there's already an app for that!

13

REGULATION AND COMPLIANCE

A h, farming...rolling hills, peaceful bird calls and...a pile of paperwork so high you need a John Deere to shift it. How hard can it be, you wonder, to grow some fruit, vegies and eggs and sell them? Not too hard. But how hard can it be to grow and sell them legally? That's another matter altogether! Welcome to legislation, licenses and never-ending rolls of red tape...rolls so long you could probably triple fence your farm with them.

If you are treating your farm as a business – or even if you're a hobby farmer selling a bit of produce here and there – it's important to understand current federal, state and local compliance laws. Unfortunately, it seems that bureaucrats feel the need to make changes to farmers' livelihoods on a regular basis, so 'current' is the key word here. If you don't work within the law, you put yourself at risk of the consequences, but as in any industry, there are some micro-farmers who keep their heads down and try to work on the fringes of the system rather than meet all the extra costs (in terms of both time and money) of compliance.

Micro and small businesses are the worst affected by regulation in Australia because while big businesses can afford to wear the costs of compliance due to economies of scale, it can take a huge chunk out of the potential profitability – and fun – of a smaller business. Even something as simple as going to a farmers' market requires you to have an insurance policy and the correct licenses for what you are selling.

Depending on what you are farming, and how you are selling it, you'll need to consider how to comply with all the regulations to work out if it's worthwhile. The foolowing are examples.

- Legislation: what are the barriers to producing what you want to produce? What licenses, standards, development applications, protocols, workplace health and safety measures and fees will you need to stump up?
- Food risk: government agencies are most concerned about consumer exposure to what they deem 'high-risk' food areas. If you are looking to sell meat, dairy, eggs or seafood, be prepared to come under extra scrutiny.
- Insurance: public liability, property insurance – it all adds up.
- Biosecurity: new legislation is being introduced all the time regarding biosecurity, and the expectations on farmers are ever-increasing. Bird flu, foot and mouth disease, Panama virus, varroa mite...put them all together and it sounds like an agricultural dystopian horror flick. If you don't want to provide front-row seats for pests and diseases on your land – or incur the wrath of your neighbours and the government – you'll need a farm biosecurity plan. It's hard to stop a bat or wild duck from flying overhead, but it's easier to control the entry of weed seeds, pests and diseases that can enter your property through vehicle and machinery movements, livestock, supplies and human visitors. From a sign posted at your front gate requesting that people call you before entering to monitoring and recording inputs such as feed and isolating new animals for 21 days, there are many things you can, and are expected to do, to stop problems spreading.

Farmers need to be across Australian consumer laws, workplace health and safety laws, land use laws, pollution laws, chemical handling laws, and state and local government laws. Links to many of these are available at www.smallfarmsuccess.com.au.

14

FOOD STANDARDS AND LABELLING

F ood Standards Australia New Zealand (FSANZ) is a government body charged with developing food standards across the two countries. What they say goes, and the Food Standards Code is then enforced by the states, local councils and other agencies.

The Food Standards Code covers these areas.

- Labelling laws and information requirements (such as a statement of ingredients, date marking of food for sale, nutrition information requirements, directions for use and storage, and country of origin labelling).
- Substances added to or present in food (such as vitamins, minerals, sweeteners and processing aids).
- Contaminants and residues.
- Foods requiring pre-market clearance such as novel foods, genetically modified foods and irradiated foods.
- Microbiological limits and meat processing requirements.
- The definition of individual foods and the standards that apply to them.

You need to know the laws relating to food to stop you getting yourself in a pickle! For example, food sold as jam, must be jam, not marmalade, and it can contain no less than 650 g/kg of water-soluble solids. Additionally, if a fruit is going to appear in the labelling for the jam, it must be made from no less than 400 g/kg of that fruit.

Local Councils often work with State bodies and may have responsibility for food safety programs and answer enquiries and enforce standards. They also inspect, assess and license food premises, including temporary market and event stalls.

Food labelling

So...have you got that degree in food labelling yet? Because you might need it! Did you know that it's not okay to just put your farm name and website address on packaged food; that you need to list a physical address? And that's just the start. Here's an introduction to some of the ins and outs of food labelling for farmers.

Country-of-origin labelling

From 1 July 2018, businesses in Australia that sell food through retail stores must comply with the Australian Government's Country of Origin label program. If the following describes your produce, your label needs to have the kangaroo symbol, some text and a bar chart showing the percentage of Australian ingredients is needed.

- Grown in Australia – for food that is 100% grown in Australia.
- Product of Australia – for all food where all the ingredients are Australian and all the major processing has been done here.
- Made in Australia – for food where the ingredients come from Australia or overseas and the major processing has been done here.

It seems naive now, but when I started in farming I used to cut out cardboard tags and write the information on them...that was until I read up on food labelling laws and realised that I'd end up with RSI if I kept trying to write the massive amount of detail that is required on a food label.

Labelling laws are set by FSANZ. They are there to protect and inform consumers. But it's not just about listing allergens (the key ones being peanuts, tree nuts, milk, eggs, sesame seeds, fish, shellfish, soy and wheat); it's about letting customers know so much more.

Although labelling standards are set by the Food Standards Code, the information on labels is also subject to fair trading laws, which means that false, misleading and deceptive claims are illegal.

The key things you need to know about the FSANZ Code, as outlined by the NSW Food Authority are summarised below.

Ingredient lists

Ingredients must be listed in descending order (by ingoing weight). This means that when the food was manufactured, the first ingredient listed contributed the largest amount and the last ingredient listed contributed the least. For example, if sugar is listed near the start of the list, the product contains a greater proportion of this ingredient.

If the product contains added water, it must be listed in the ingredient list according to its ingoing weight, with an allowance made for any water lost during processing, e.g., water lost as steam. The only exceptions are when the added water:

- makes up less than 5% of the finished product
- is part of a broth, brine or syrup that is listed in the ingredient list
- is used to reconstitute dehydrated ingredients.

Sometimes, compound ingredients are used in a food. A compound ingredient is an ingredient made up of two or more other ingredients, e.g., canned spaghetti in tomato sauce, where the spaghetti is made up of flour, egg and water. All the ingredients that make up a compound ingredient must be declared in the ingredient list, except when the compound ingredient is used in amounts that constitute less than 5% of the final food. An example of a compound ingredient that could be less than 5% of the final food is the tomato sauce (consisting of tomatoes, capsicum, onions, water and herbs) on a frozen pizza.

However, if an ingredient that makes up a compound ingredient is a known allergen, it must be declared, regardless of how little is used.

Percentage labelling

Most packaged foods must carry labels showing the percentage of the key or characterising ingredients or components in the food. This allows you to compare similar products.

The characterising ingredient for strawberry yoghurt would be strawberries, and the label would say, for example, 9% strawberries. An example of a characterising component could be the cocoa solids in chocolate. Some foods, such as white bread or cheese, may have no characterising ingredients or characterising components.

Once you have your ingredients list prepared, it's time to make use of the...

Nutrition Panel Calculator

So, your label is filling up with text fast, but you still need to save space for the legally required (for most packaged foods) nutrition panel. If you're not a math genius, use the FSANZ's Nutrition Panel Calculator (NPC) to calculate the average nutrient content of your food.

The calculator is relatively easy to use. You search the NPC database for the ingredients in your product, e.g., strawberries, sugar, lemon juice, and then type in the amount of each ingredient you used and the final batch weight of the product/recipe (this is normally different to the total of the ingredients you used, as weight can be lost to steam etc.). You also type in the serving size and the number of serves per package and the calculator then generates the nutrition panel so you can print or save it.

What information must be included in the nutrition panel?

(a) the number of servings in the package, expressed as either:
    (i) the number of servings of the food
    (ii) if the weight or the volume of the food as packaged is variable –

the number of servings of the food per kilogram, or other unit as appropriate

(b) the *average quantity of the food in a serving expressed in:

(i) for a solid or semi-solid food – grams

(ii) for a beverage or other liquid food – millilitres

(c) the *unit quantity of the food

(d) for a serving of the food and a unit quantity of the food:

(i) the *average energy content expressed in kilojoules or both in kilojoules and in kilocalories; and

(ii) the average quantity of

(A) protein, carbohydrate, sugars, fat; and

(B) subject to subsection (4), saturated fatty acids, expressed in grams; and

(iii) the average quantity of sodium, expressed in milligrams or both milligrams and millimoles; and

(iv) the name and the average quantity of any other nutrient or *biologically active substance in respect of which a *claim requiring nutrition information is made, expressed in grams, milligrams, micrograms or other units as appropriate; (e.g. if a claim regarding nutrition information is made in respect of:

(a) cholesterol; or

(b) *saturated,* trans, *polyunsaturated or *monounsaturated fatty acids; or

(c) omega-3, omega-6 or omega-9 fatty acids;

a nutrition information panel must include declarations of the trans, polyunsaturated and monounsaturated fatty acids

(d) *trans fatty acids.

Read the entire code for more information, as there are sections relating to claims in respect of dietary fibre, sugars or carbohydrates, phytosterols, phytostanols or their esters, lactose, omega-3 fatty acids, salt or sodium and potassium. There are also different ways to express elements in the nutrition panel, for example, the word 'serving' may be replaced by 'slice', 'pack' or 'package'.

WEIGHTS AND MEASURES

Hushed whispers went from market stall to market stall and finally reached me, "the inspector guy is here!"

"What inspector guy?" I asked, "the council food inspector?"

"No! Trades and Measures!"

Having never been inspected before, I had no idea what to expect as he circled my stall, lifting jars and checking labels. We had a nice chat, he said I'd done an appropriate job regarding measurement, and I watched as he moved on to the next stall, where he spent a long time investigating their electric scales and signage. Meanwhile, some other stallholders were throwing produce and cartons full of bottles back into their vehicles. Why? Because they knew they hadn't complied perfectly with the law.

The National Measurement Institute (NMI) is a division within the Australian Government's Department of Industry, Innovation and Science, and, you guessed it, they are the experts on measurement. They've provided the following information to help market and roadside stall operators understand their obligations to customers.

Disclaimer: *"The purpose of this publication – Guide for Market Stall-holders and Roadside Traders – is to provide you with general information only and should not be relied upon for any legal, business or personal purpose.*

Nothing in this communication shall be taken in any way to replace the provisions of the National Measurement Act 1960 (Cth), the National Trade Measurement Regulations 2009 (Cth) or any other legislative instruments made pursuant to the National Measurement Act 1960."

What is trade measurement?

Trade measurement is the system that controls the buying or selling of any good or service where the value is determined by a measurement. Whether you are selling cheese by weight, milk by volume, or fabric by length, it is important as a business to know your trade measurement obligations.

Who benefits from trade measurement?

Everyone benefits when correct measurement is applied in trade. Consumers benefit from getting the goods or services they pay for. Businesses benefit by reducing potential oversupply. The economy benefits from consumer confidence through a consistent and reliable system.

What if I use measurement to sell goods?

If you sell goods by measurement (e.g. weight, length or volume) you will need to have a measuring instrument such as a weighing instrument (scale). Measuring instruments used in trade must be of an approved design or 'pattern' and be marked with an NMI approval number or a National Standards Commission (NSC) approval number. An example could be NMI 6/4D/355 or NSC 6/4D/220. Approved measuring instruments must also be verified by a servicing licensee or inspector before they can be used for trade.

If you use a measuring instrument in trade, you must make sure:

- it is an approved type
- it has been verified before use
- it indicates zero before use

- it is level when in use
- it is positioned so that the customer can easily see the measurement process
- it is kept clean and in good working order
- it is verified after each repair or metrological adjustment
- it is suitable for its intended purpose
- it is used in the correct manner.
- it is checked regularly.

What if I use a measuring instrument to sell goods?

If you sell fruit, vegetables or some other food by weight, the scales that you use must be verified by a servicing licensee or an inspector.

When selling a quantity of a good by weight, the weight of any packaging (the 'tare') should not be included. This process is called 'taring off'. Charging a customer for the weight of a good without subtracting the weight of the packaging is an offence and could result in a costly financial penalty.

You can write the weight on the packet by hand rather than pre-printing labels or bags, but only when the produce is packed and sold on the same premises. In all other cases, the weight must be printed or stamped.

What if I sell goods by count?

If you sell a good by count, e.g., a dozen oranges or 100 nails, the packaging must be marked with the accurate count. The only time that you do not have to mark the count on a pack is when the produce is packed:

- in a clear transparent bag or package and
- there are less than nine articles in the pack.

You can write the count on the packet by hand rather than pre-printing labels or bags, but only when the produce is packed and sold on the same premises. In all other cases, the count must be printed or stamped on the packaging.

What if I sell pre-packaged goods?

All goods pre-packed for sale must be marked with the net measurement (that is the weight or number without the packaging) on the principal display panel. This panel is the part of a package that is most likely to be shown under normal conditions of display.

The measurement statement must be:

- clear to read, 2 mm from the edge of the principal display panel and at least 2 mm from other graphics
- in the same direction as the brand or product name and in a different colour than the background.

All pre-packages (unless specifically exempt) must be clearly marked with the name and street address of the packer or party accepting responsibility for packing. The use of a post office box, locked bag, telephone number or email is not acceptable.

Pre-packages must not contain less than the stated amount at all times prior to sale.

If the article is likely to lose weight or volume over time, through evaporation, dehydration or other means, the packer must make allowances for any expected losses in the measurement when packaging the product.

You are only responsible for the measurement statement on goods that you have packed yourself.

16

ANIMAL WELFARE

ccording to the RSPCA, there are no national laws applying to animal welfare, but all states and territories regulate animal welfare in their jurisdiction. This means that in addition to your own considered ethical treatment of animals, you need to be aware of the current legislation in your state.

As an example, the purpose of Queensland's Animal Care & Protection Act 2001 is to:

(a) promote the responsible care and use of animals

(b) provide standards for the care and use of animals that:

- (i) achieve a reasonable balance between the welfare of animals and the interests of persons whose livelihood is dependent on animals; and
- (ii) allow for the effect of advancements in scientific knowledge about animal biology and changes in community expectations about practices involving animals

(c) protect animals from unjustifiable, unnecessary or unreasonable pain

(d) ensure the use of animals for scientific purposes is accountable, open and responsible. It specifies that a person in charge of an animal (either through owning, leasing, licensing or other proprietary interest; and/or has custody of, and/or is employing or has engaged someone else who has the custody of the animal and the custody is within the scope of the employment or engagement) owes a duty of care to it.

This duty of care means that they need to:

(a) provide the animal's needs for the following in a way that is appropriate:

(i) food and water

(ii) accommodation or living conditions for the animal

(iii) to display normal patterns of behaviour

(iv) the treatment of disease or injury; or

(b) ensure any handling of the animal by the person, or caused by the person, is appropriate.

The Act prohibits cruelty to animals. A person is taken to be cruel to an animal if the person does any of the following to the animal:

(a) causes it pain that, in the circumstances, is unjustifiable, unnecessary or unreasonable

(b) beats it so as to cause the animal pain

(c) abuses, terrifies, torments or worries it

(d) overdrives, overrides or overworks it

(e) uses on the animal an electrical device prescribed under a regulation

(f) confines or transports it:

(i) without appropriate preparation, including, for example, appropriate food, rest, shelter or water; or

(ii) when it is unfit for the confinement or transport; or

(iii) in a way that is inappropriate for the animal's welfare; or

Examples for subparagraph (iii) –

• placing the animal, during the confinement or transport, with too few or too many other animals or with a species of animal with which it is incompatible

• not providing the animal with appropriate spells

(iv) in an unsuitable container or vehicle

(g) kills it in a way that:

(i) is inhumane; or

(ii) causes it not to die quickly; or

(iii) causes it to die in unreasonable pain

(h) unjustifiably, unnecessarily or unreasonably:

(i) injures or wounds it; or

(ii) overcrowds or overloads it.

A person in charge of an animal must also not abandon or release it (or leave it for an unreasonable period) unless the person has a reasonable excuse or the abandonment or release is authorised by law.

There are specific animal welfare issues that, although they may not be currently legislated against, are not ideal. You should also look into industry-led codes of conduct, and at the other end of the scale, suggested codes of conduct by animal welfare groups, to see how you can offer best practice and care for your livestock. Animal welfare is only going to become a hotter topic for key groups of customers.

17

LIVESTOCK TRACKING AND PROPERTY IDENTIFICATION

Since 2012, anyone who keeps or owns livestock – even if it's just one old cow or goat – is required to have a Property Identification Code (PIC). This unique eight-character alphanumeric code is allocated by various states and territories to parcels of land used for agriculture.

According to Agriculture Victoria, "The purpose of property identification is partly for tracing and controlling disease and residue problems that may be detected after leaving the farm, but also for locating properties and owners that have livestock when an outbreak of a disease that may threaten their enterprise is detected in an area. Traceability systems provide confidence to consumers in domestic and overseas markets that the products they buy are of good quality."

If you graze any of the following livestock on your property, you require a PIC (or if you are agisting animals on your or someone else's property, the property will require a PIC):

- one or more cattle, sheep, goats, pigs, camelids (alpacas and llamas), deer, bison, buffalo, equines (horses and donkeys)
- more than 100 poultry (i.e. domesticated fowl, chickens, ducks, geese, turkey, guinea fowl, pigeons, quail or pheasants) or 10 emus or ostriches.

Additionally, all livestock businesses (sale yards, cattle sales, abattoirs, knackeries and stock agents) must have a livestock PIC.

There is even a plant PIC in some instances. For example, individuals who grow the following prescribed horticultural crops within Victoria must have a plant PIC:

- 0.5 hectares or more of grapevines
- 20 or more chestnut trees.

To apply for a PIC, you need to contact your state agriculture department to find out which agency you need to liaise with. For example, in NSW it is the Local Land Services department. The PIC in NSW is usually renewed every three years.

The NSW Department of Primary Industries states that "Every livestock producer should have a PIC because:

- a PIC is printed on approved livestock identifiers (NLIS tags and devices)
- the NLIS database records 'To' and 'From' PICs for every individual cattle movement and every mob-based movement for sheep and goats
- a 'From' PIC is recorded on stock movement documents such as National Vendor Declarations (NVD) and Transported Stock Statements (TSS)
- a PIC is required for industry quality assurance programs, e.g., Livestock Production Assurance (LPA) and Australian Pork Industry Quality (APIQ) programs
- an NVD or PigPass NVD is required for each consignment of stock for sale or slaughter. Owners of small farms that have not previously been assigned a PIC are encouraged to obtain a PIC for their property.

PICs also form the basis of the National Livestock Identification System (NLIS) as they provide traceability to specific properties."

National Livestock Identification System (NLIS)

The NLIS involves tagging the ear of cattle with an approved electronic tag or inserting a device into rumen boluses. These devices can then be read by an electronic reader to identify the origin and movement of the cattle between properties with different PICs. Movements are recorded on a centralised national database.

NLIS Cattle was introduced in NSW on 1 July 2004, and according to the NSW DPI "The National Livestock Identification System (NLIS) is Australia's scheme for the identification and tracing of livestock. This system enhances Australia's ability to respond quickly to a major food safety or disease incident in order that access to key export markets is maintained. It is a key industry initiative in partnership with governments across Australia."

On January 1 2006 the NLIS Sheep & Goats was also introduced. It uses visually readable (rather than electronic) ear tags printed with a PIC. A lot of paperwork also needs to be filled out when moving or selling sheep and goats. The owner is required to declare the location where the sheep were bred on a National Vendor Declaration (NVD) form. These requirements also apply to any individual or company who trades livestock, including swapping or selling animals at sale yards or through other informal channels such as neighbours, friends, markets and websites. Random checks are carried out to ensure that livestock owners are meeting their obligations.

18

WORKPLACE HEALTH AND SAFETY

S o, you want to be a farmer, right? But hopefully you don't want to become a statistic. That's why taking farm safety seriously and being aware of and acting according to safe work legislation and practices are essential.

According to Agriculture Victoria, farmers and farm workers are more likely to be seriously injured or die at work than other Victorians... and not just by a small margin. Although the farming industry employs only about 3% of the Victorian workforce, it accounts for approximately one-third of workplace deaths across the State. That's a startling and sobering statistic.

Why can farming be so dangerous? Let's look at some of the risks.

- Farmers often work alone or on remote parts of the property.
- Farmers work long hours, and this can lead to fatigue...and therefore accidents.
- The work can involve lifting large, heavy, bulky and awkward items (if you've ever tried lifting an entangled ram out of a fence you'll have a first-hand understanding of this).
- There is repetitive strain from doing the same job multiple times...whether it's weeding, pulling, tugging or pushing (if

you've ever tried moving a stubborn billy goat without a bucket of food you will also understand this).

- Tractors are large, heavy vehicles that can roll and run over and into things.
- Power-driven attachments on tractors can be very dangerous, be it the moving parts themselves or the effect they have on the surrounding area, e.g., a slasher can send a spear of wood or a star picket flying at high speed into soft flesh.
- Unguarded augers can trap body parts.
- Quad bikes, motorbikes, boats and other farm vehicles, machinery and tools can cause injury and death.
- Tanks, troughs, dams, creeks, rivers, oceans and flooded crossings can lead to drowning incidents.
- Animals can crush, ram, kick, throw, bite and generally be unpredictable.
- Guns on properties can lead to accidents and misadventure.
- There is also potential exposure to sun, heat, cold, excessive noise, chemicals, dusts, allergens and pathogens.
- Cuts, falls, infections, slips and trips (yep...any farmer will experience some of these!).
- Fatigue.
- Depression.

With so many risks, you are legally responsible for the safety of not just yourself and your family, but also any contractors, visitors, neighbours and farm workers who enter your property. Workers need to be trained and you need to minimise risks wherever you can to ensure a safe workplace.

Each state provides guidelines and legislation to ensure that you know your responsibilities, such as the following.

- Ensuring hearing protection is used when operating machinery.
- Providing adequate respiratory protection.
- Wearing a helmet when riding motorbikes or quad bikes (and attending safe driver training).

- Appropriate storage, training in and application of chemicals (or use of safer alternatives).
- Signage to warn of risks, hazards and protocols.
- Implementing control strategies for fatigue such as including shorter shifts, rotating staff, or introducing rest periods.
- Securely fencing the house yard and supervising children.
- Not operating machinery under the influence of alcohol.

By facing risks head on, doing a safety audit for your farm and farming practices, and implementing sensible strategies, it's easy to minimise the risk you, your family, staff and visitors face.

FARMER SPOTLIGHT: Oysters

A passion for water and an ideal location cracked open a new farming venture for this family.

Farmers: Casey and Susan Lowick
Farming: Sydney Rock oysters
Farm name: Blue Coast Oyster Company
Farm size: 6 ha of water (14.83 acres)
Average annual rainfall: 1 225 mm
Where: Forster-Tuncurry, NSW
Farming since: 2012

On a family holiday in Tuncurry, swimming-pool builder and experienced farrier Casey rose early one morning to be with his sleepless toddler son. While his wife Susan (a photographer) stayed snoozing, he and his son watched the oyster fishermen taking to their boats, gliding out at sunrise across the aqua waters of the lake. Right there and then Casey decided that water-based farming was what he wanted to do. The couple, who had been keen to start their own business, also saw oystering as a way to leave something for their son when they were gone.

To prepare for their oyster farming venture, the Lowicks didn't undergo formal training, but did talk to lots of other farmers, scoured the internet, and watched YouTube videos to learn the ins and outs of oysters. They believe that being inventive, resilient and resolute are three key personality traits needed by farmers, and these traits need to be backed up by a strong work ethic, physical strength, openness to learning and good business sense.

"Oyster farming can be an amazing and somewhat scary life," said Casey. "There are beautiful things to see all day – dolphins, fish, sunrises and sunsets, storms, flat water, rough water – but you always need to be on your toes because you can never be too careful on the water."

The Lowicks set up their venture in the Great Lakes region of NSW because it offers the cleanest water to farm oysters in all of Australia, and some of the cleanest in the world. Working 60+ hours per week (95% on physical labour and 5% on admin/management), they sell their oysters via wholesale and retail outlets and to chefs. The Lowicks successfully sold part of their business in 2017, retaining a few leases on which they are implementing new techniques.

"Setting up the venture was a big risk and I was working 80+ hours a week for the first few years to survive," said Casey. "But it's been so worth it. I get to work outside, on the water, in a beautiful climate, creating jobs, a true delicacy of a food, and a future for my family. I also get to indulge my inventive nature and am working on some projects that might revolutionise how oysters are farmed."

Food for thought from Casey and Susan

What are the basics needed for oyster farming?
A background in any type of farming or fishing would be ideal, but to just be inventive and open minded. You need a boat licence and an aquaculture licence to farm oysters, along with a land based depot to process them.

What are the threats to the industry?
Flooding can close the river/lake so we can't harvest, and disease is always a threat.

Opportunities in the industry?

Many older oyster farmers are retiring and there is room for new farmers.

What makes a good farmer?
Susan: someone who creates food and jobs without leaving a large footprint. Casey: someone who is sustainable, leaving the land – and water – in a better condition than when they started. In the oyster business, it's someone who can read the tides and the weather, and successfully grow healthy, fat oysters.

Best piece of advice from another farmer?
Don't trust the middlemen (oyster processers) unless you've had a long relationship with them and get paid up front.

Worst piece of advice from another farmer?
"Buy a lot of leases to grow your oysters." The Lowicks realised early on that quality over quantity was a better approach for them and this saved them thousands of dollars.

What would you farm if you weren't involved in your current activity?
Cattle; the mid-north coast is a great region for them.

Casey and Susan's top 5 tips
1. Be willing to work hard.
2. Have a second job to make ends meet for a couple of years.
3. Understand the risks of farming.
4. Always be thinking of new and better ways of doing things.
5. Just do it – it's worth every second of the hard work.

What's the best way to enjoy your product?
Fresh and raw!

PART IV

PART IV

19

MARKETING

No matter what you are producing or how well you are producing it, you still need customers.

Imagine a scenario in which what you've planted and cared for has now grown to maturity. It's grown so well in fact that it's not until you've picked your first ten thousand bulbs of garlic and twenty thousand limes that you begin to seriously consider how you're going to sell it all, and how you'll get the best price.

Please, don't let that be you. Don't be the farmer who hasn't planned a sales and marketing strategy, and so ends up with mummified garlic still hanging from the rafters years later. Don't be the farmer who through lack of planning is forced to sell the truckload of limes at the lowest possible price. Don't be the farmer who ends up with fruit rotting on trees and wins the local fruitfly breeding championship by a country mile because it's not worth the effort to pick it. Be the farmer who farms in a business-like manner, has a plan and is focussed yet flexible. Be the farmer who has a future.

It can be difficult to imagine the tiny seedlings you're planting now growing to maturity, their branches laden with nuts. It can be hard to grasp how many cuts of meat and 500 gram packets of mince you can produce from a small herd of Angus. Even finding a market for the freshest of eggs once you've exhausted the purchasing power of your neighbours, family, friends and colleagues can be difficult. The good news is that it doesn't have to be that way if you plan for it. And marketing can be fun to do!

That's what this next section is about. It's about finding your customers and getting the best price for your produce in the most efficient way possible. It's about building your farm's brand and building relationships.

Your brand

We're not talking about the initials of a farm being burnt into a thoroughbred's hide. We're talking about the feeling customers get when they think about your business and produce. Your brand is what is seared into a customer's mind about you, and it's important because it can help you increase sales, improve loyalty and get a better a price for your produce.

We all have a personal brand amongst our friends and family. We might be known for being calm and trustworthy, or hardworking but volatile. We might be the life of the party, the shoulder to cry on, the peacemaker, the restless spirit or the meticulous planner. But businesses also have brands, and as a farm your brand sums up what consumers think of who you are, what you do, how you do it, your qualities, your trustworthiness and your value. Everything you do as a farmer, business owner, marketer and neighbour builds customers' perceptions of you, and therefore your brand. Think about the following examples.

- Marketing materials – your logo, font, business card, website, social media presence, advertisements, product photography, vehicle and other signage, product labels, media interviews and packaging.
- Customer contact – how you answer the phone, the tone of your emails, and how you deal face to face with customers and with post-sale service and support (including handling compliments and complaints).
- Industry contact – how you come across to suppliers, competitors, and industry and government organisations.
- Your product – its quality, its presentation, how useful or

unique it is, and how well it fulfils the needs of your customer.

When you think of Qantas, do you think of a budget airline or safety? When you think of Red Bull, do you think of being bored or getting a shot of adrenalin? Every major company in the world spends money and time developing and protecting their brand because they know its value when it comes to getting customers to trust and choose them over the opposition. Whatever you're selling, you're going to need to invest time and thought into your brand.

So, how do you want customers to perceive your farming business and produce? This depends on who your core target customers are, because your brand needs to directly engage their core values. Let's look at the different focusses of each of two different types of carrot farmers.

Farmer Red produces 10 000 tonnes of carrots a year. Target customers are wholesalers. These customers are seeking a reliable and consistent supply of uniform carrots. Therefore, Farmer Red approaches everything in a consistent manner. His cartons and trucks are clearly labelled, with no fancy fonts. He and his staff have a direct but friendly, no-fuss approach when dealing over the phone and in person with wholesalers. He invests in the latest precision grading and sizing machinery to back up the customers' need for reliability and uniformity. Farmer Red and his workers have a simple logo emblazoned on their workwear and a simple one page website. Other wholesalers want to do business with him because of his solid reputation, but Farmer Red stays loyal to his long-time customers. Farmer Red reeks of reliability, and that's exactly what his customers need and appreciate.

Meanwhile, Farmer Orange produces 100 tonnes of heritage variety carrots a year. Target markets are direct customers and retailers. Her farmers' market customers, seeking a connection with food and growers, like that her carrots are different shapes and sizes, as it adds to her authenticity as a small grower. She wears her hair in pigtails and drives a bright orange Kombi Van emblazoned with her funky logo. Farmer Orange also value adds her carrots into jars of Vietnamese daikon. She also pickles them, ferments them with cabbage and makes a carrot kombucha. She packages these products in fat glass jars adorned with her funky logo and they sell at a premium price on her stall, as well as at health food shops and upmarket delicatessens. Farmer Orange comes across as bright, creative and passionate. She does artistic social media posts and sends out an email newsletter once a fortnight to her customers (direct and retail) with carrot recipe ideas. People enjoy her products not just because of how they taste and the perceived health benefits, but also because they like being associated with such a quirky, earthy person. Farmer Orange puts life into carrots, and her customers want to put that life into their own bodies by eating her carrots. They're also going to tell their friends.

Your brand is your company's reputation. It's unique, and it will provide support to, or detract from, your marketing activities and, ultimately, your financial bottom line.

Web presence strategy

A decade ago, the heading here would have been '*website* strategy', but it's now called a web *presence* strategy because your farm's online life might need to be a lot more than just a page or two. An online strategy helps you reach, inspire, network with, inform, convert and develop relationships with your customers and peers. The following areas should all be part of a coordinated focus.

Website

Build it yourself using a free or economical option like WordPress, Wix, Duda, Squarespace or Weebly, or outsource the development to a web company. Whatever you do, make sure it's mobile friendly! By having your own website, you have the means to project your unique identity and information. Use lots of luscious photos of your product growing on the farm, and show the ways in which it can be used. A picture of you, the farmer, is also great, as it shows you are proud of what you are doing.

Key-word marketing

From your distinct URL to every line of text and photo caption, your website will be more easily found if you pay attention to the key words people usually use to search for products like yours. Key-word marketing is a science, but it's easy to grasp the basics by researching it on the internet.

Social media

Facebook (especially locally based Buy, Swap & Sell sites), Instagram, Twitter, Pinterest, Tumbler and YouTube are the biggies, hosting lots of content and therefore lots of potential contact with your target audiences. You don't necessarily need a presence on these platforms, but ignore Facebook in particular at your peril. It has proven to be an excellent connector, and although there will always be the next big thing targeting different generations (such as Snapchat), Facebook and Instagram are too big to ignore. That said, not being on them won't break your farming venture; it's just that it won't get a lift from them.

A word of caution: it can be easy to end up spending your day looking at what everyone else is doing online, rather than making more of your own backyard – try not to get sucked in to wasting too much time netsurfing.

Permission marketing

By getting permission from people to send them information, you can create email lists. You can get permission offline by having people fill in a form, or online by having a sign-up button. This database then becomes your way to contact an interested audience when you have a special offer or other information to share. Don't overwhelm people with too many emails, but a regular monthly or seasonal outreach where you share information such as recipes, news articles and farm news is a way of developing a bond with your customers, and potentially selling more to them.

Review sites

Business review sites from TripAdvisor to Google to Facebook pages mean that anyone can comment on your products. It can be great when you receive positive reviews and heartbreaking when someone disses your produce (and therefore your efforts). Many customers look to reviews to give them the confidence to buy from you, so it's important to have a presence if you are dealing directly with consumers.

Blogs, vlogs and podcasts

Blogs (written), vlogs (audiovisual) and podcasts (audio) are not for the faint-hearted as they will take up a lot of your time and headspace! They also require a long term commitment to get results. A website and social media strategy that involves posting updates, news and images might be easier to handle than a stand-alone blog unless you absolutely love communicating and can always find something to say. Over time, a blog can help people to find you through search engines, but you could also post "Latest News" updates on your stand-alone website. WordPress offers a combination of both.

Online aggregators

Online aggregators bring multiple producers together on the one site. They market extensively and use the collective attraction of so many products to entice customers to buy. We'll cover online aggregators in more detail in the retail section.

The internet offers modern farmers a great way to connect with customers, and it's easy to start with a simple website and grow and learn from there. Promote your URL on product labels, your car door, and in any advertising so that you can draw customers to your website to sign up for future permission marketing activities.

More marketing tips

In addition to your focus on your brand and your web presence, there are all sorts of simple marketing techniques you can employ to help grow your customer base, including the following.

- **Media relations.** Raise your profile locally and further afield by getting in touch with the media with good news stories, as

well as information and commentary they might find useful
or controversial. Small publications in particular love great
photos to accompany stories, so take high-resolution photos
that are good enough for print-based media as well as
the web.

- **Images.** Speaking of photos, unless you are a great
  photographer, get a professional in so you have a great stock
  of photos you can use for your website, media releases and
  brochures.
- **Signage.** Have your road vehicle sign-written or use magnetic
  signs on your car doors and always drive courteously.
- **Advertising.** The cheapest and most focussed advertising you
  can do is via the internet, where you can target your audience
  right down to their interest in animal rights through to
  cooking through to whether they have children with allergies.
  Internet advertising is cheap compared with traditional
  media in that it lets you test which ads work and which don't.
  That said, a large, repeated campaign in local media (radio,
  cinema, TV, billboard and print) will have an effect, but it just
  might be more scattergun.
- **Trade shows and consumer events.** Take out a stand to
  showcase your produce to potential wholesalers, exporters,
  foodservice professionals and consumers.
- **Awards.** Enter awards ranging from local business awards to
  food product awards. When we won various local, state and
  national tourism awards, it not only generated publicity for
  our farm but also gave us accolades we could place on our
  website to give customers that extra level of trust and
  motivation to book with us.
- **Labels.** Ensure your product labels have a call to action or
  point people to a website where they can learn more about
  your farm.
- **Landmarks.** Visually attract attention to your property with
  an art installation of some sort in your roadside paddock.
  This could be done through landscaping, topiary, the
  addition of a large wire sculpture or old farm equipment
  placed atop mounds.

- **Track and measure.** How will you know which form of marketing is working if you don't ask people how they heard about you? By tracking and measuring what works and what doesn't, and changing your promotional efforts to maximise the effect, your marketing budget will go a lot further.
- **Gather a tribe.** Invite people to join your mailing list so that over time, you can create a database of customers you can contact directly with offers, information and opportunities.

Sales and product strategy

When you investigate the various ways to sell your product, keep in mind the following factors.

- **The scale and specialty of your enterprise.** How much will you need to sell? Will it be 800 tonnes of wheat or 10 tonnes of spelt? Will it be 4 000 kg of mixed blossom honey or 4 000 small jars of medicinal honey? Will it be 15 000 kg of potatoes or 5 000 bottles of potato vodka?
- **The stability of your produce.** How fast will it need to be sold to customers? This relates to shelf life. Fresh-cut herbs and flowers need to be marketed and sold quickly before they wilt, so you need your customers and logistics lined up flawlessly. Milk needs to be sold within days, whereas cheeses might be fine for 12 months (and some are even better when they're aged for longer). Processed olives and vinegars can have a shelf life of years.
- **The farm's location.** Are you close to major population centres? How easy and economical is it to transport your produce to customers? What will be the cost of freight, couriers or postage? Is it likely that enough people will come to the farm to buy your produce, and even if it is, do you really want that?
- **Your skill set.** Are you a natural salesperson who enjoys being around people, or would you prefer to stay on the farm and leave the selling of the produce to someone else?

- **Margins.** What margin do you need to be profitable? Can you get the margin you need from a wholesaler or co-op, or will you also need to investigate selling direct to customers and/or value adding your produce to get the price you need? Is your pricing sustainable? For example, at that price, can you increase capacity? Can you handle the cost of a key ingredient rising by 20%? Can you keep the product at that price if you have to pay staff to do the work rather than you doing it?

Now drill down a bit more. What type of product are you selling? Is it a commodity, unique, artisanal, fresh, frozen, organic or biodynamic? Does it have a stable or long shelf life? What is its size, weight and bulk? Will it need to be refrigerated for shipping? Is it sold as is, or value added? How does it need to be packaged and labelled?

How is your product used? Is it:

- **a staple in the pantry of most homes**, e.g. dried herbs and spices, tea, olive oil, vinegar?
- **a staple in fridges**, e.g. butter, milk, salad leaves, condiments?
- **consumed daily**, e.g. fruit, vegetables, dairy, meat?
- **enjoyed once or twice a week**, e.g. haloumi or salmon?
- **snapped up during certain seasons**, e.g. cherries?
- **for lazy weekend consumption**, e.g. free-range bacon and eggs?
- **a spur-of-the-moment treat?**
- **to impress guests** at a dinner party, restaurant or function, e.g. micro-herbs, edible flowers or specialty cuts of meat?
- **for presentation as a gift**, a thank you or a peace offering?
- **for decoration?**
- **a one-off specialty item?**

Who might be your customer (keeping in mind that you don't want everyone to be your customer – there are good ones and bad ones)? Good customers for your farming business might include:

- family meal preparers

- pensioners
- foodies
- browsers/spur-of-the-moment buyers
- gift buyers
- retailers
- chefs, caterers or food-service operators
- wholesalers
- exporters.

Many farmers who are familiar with the saying "don't put all your eggs in one basket" sell through multiple channels. Investigate which ones you identify with, or what mix might be best for your enterprise. The main ones, which are explained in more detail in the following pages, are as follows:

- a middleman – wholesaler
- food services (chefs and catering operations)
- retail outlets
- a farmgate outlet (this can be staffed or unstaffed)
- your own retail outlet
- direct-to-customer database (marketing)
- online through your own store
- online through an aggregator
- farmers' markets
- community supported agriculture (CSA).

Cooperatives and collaborative farming groups

Cooperatives bring together growers of similar products in an effort to achieve economies of scale and the best price for the produce. Co-ops charge member farmers a levy based on what the farm produces. For example, a mango farmer might pay an amount based on how many kilograms or trays of mangoes are produced, while a pig farmer might pay an amount per carcass. This levy is then used by the co-op in areas such as marketing, sales, public affairs, biosecurity, and research and development.

Well run co-ops allow the farmer to concentrate on farming, while experts at the co-op take care of research, collective bargaining, marketing, sales and potentially central processing and export facilities. A co-op should be able to negotiate better prices on the equipment its member farmers need. It might be able to apply for government funding for export assistance and develop foreign markets. It might instigate research and share the findings to help member farmers to produce better crops. It might be able to offer bulk storage to enable farmers to keep their on-farm costs down. Not all collaborative farming is in the form of co-ops, however. Some farmers simply work together in a less structured way to benefit each other, for example, by taking each other's produce to markets.

A diverse array of produce is marketed through co-ops in Australia, but it is not compulsory to join one. You should only join if it suits the way you want to do business. Prominent co-ops include The Australian Mango Industry Association, The Geraldton Fishermen's Co-operative (rock lobsters), NORCO (dairy), Batlow Fruit Cooperative (apples) and the Rice Growers Cooperative (which markets rice to consumers through the SunRice brand). Co-ops can offer real benefits to farmers located a long way from key markets. Some Australian co-ops end up being bought out by foreign buyers or become listed companies on the stock exchange, and they are not for everyone.

If you are very independent, and like doing your own marketing, the following section discusses ways of selling produce that you might find more challenging, enjoyable and profitable.

FARMER SPOTLIGHT: Saffron

A former human resources manager brings spice to Tasmania.

Farmers: Nicky and Terry Noonan
Farming: Saffron
Farm name: Tas-Saff
Farm size: 1.2 ha (3 acres)
Where: Glaziers Bay, Huon Valley, Tasmania
Average annual rainfall: 712 mm

Farming since: 1990

Food lovers Nicky and Terry Noonan left Sydney in 1989 looking for a complete lifestyle change. They purchased a small cottage on just three acres of land in a temperate, fruit-growing region an hour from Hobart. It was while making a paella for visiting friends that the dried red stigma of the saffron flower (*Crocus sativus*) implanted itself in their lives.

"It was just so difficult to track it down, and when we finally did, it was expensive," said Nicky. "We became fascinated by it."

The Noonans' research showed that most of the saffron available in Australia was imported, but that Tasmania had the ideal climatic conditions to grow it in. Nicky became intrigued by the idea of starting a new industry and replacing the inferior imported product. She researched the opportunity for five years, travelling overseas and learning from other growers, and is now one of the most successful saffron producers in the country. She has further enhanced the business, and the industry, by establishing a network of more than 50 Australian growers who supply threads for the Tas-Saff product range.

Nicky thinks all farmers need to be stubborn and unwilling to accept defeat. She needed to call on these traits herself after a one-in-a-hundred year flood ruined their crop in their third year. She also believes strong family support from her husband, son and a business mentor have had a huge impact on the farm's success.

Nicky's typical 50–60+ hour week (40% physical labour and 60% admin) involves manufacturing, marketing and sales, quality assurance, administration, accounting, new product development, crop maintenance and cleaning. The company supplies hand-picked threads to chefs, wholesalers, retailers, farmers' markets, farmgate shops, online customers and export markets.

Nicky's focus on growing a premium, high-quality product supported by good packaging is a strategy that has helped the company stay on track, even when a cheaper, unlabelled, imported product floods the market. Having a local agricultural contract farmer working alongside her with good local employees and artisan producers value adding the product also enables her to search for better export opportunities and to lobby for transparency and a level playing field for Australian farmers.

"The downsides of farming are the stress on the body as I age and not enough time off. There's also the irregular income and low superannuation. But that's made up for by living in a small town with great community spirit in one of the most beautiful parts of Australia."

Food for thought from Nicky and Terry

What are the basics required for your farming activity?
Experience in farming/agriculture (or willingness to do extensive research and trials). Also needed are sound administration, marketing and sales experience, good soil and land conditions, temperate climate, great staff, contractors and community support.

What are the threats to the industry?
We compete with cheap, imported product that has nowhere near the same labelling or quality assurance requirements imposed on the locally grown product. Wage rates in Australia make it extremely difficult to provide a competitive product with suitable profit margins for wholesalers, retailers and chefs.

Opportunities in the industry?
There are great opportunities online and in export markets where people are seeking a quality product with known heritage. Our industry also offers growers in our network the opportunity to sell their product at fair, stable prices.

What makes a good farmer?
You need to be able to adapt to new opportunities, be a good communicator and have the ability to prioritise and delegate.

Best piece of advice from another farmer?
Focus on ensuring that the climate and land conditions are suitable for the crop.

Worst piece of advice from another farmer?
What's saffron? Never heard of it! There's millions to be made in the wine industry. Grow grapes and start a vineyard!

What would you farm if you weren't involved in your current activity?
That's a hard one, but 27 years ago I thoroughly investigated olives, cherries and organic garlic.

Nicky and Terry's top 5 tips
1. Never start a project undercapitalised.
2. Really focus on land suitability/climatic conditions.
3. Sort out your logistics.
4. Find and develop market opportunities for your product.
5. Set up in an area where you can access good staff.

What's the best way to enjoy your product?
Saffron rice, saffron scones and saffron poached pears...and I love sitting back at the end of the day on the veranda overlooking the Huon River and the Hartz Mountain Range with a Tas-Saff Grower's Own saffron gin!

20

DIRECT TO CUSTOMER RETAIL

I t might make sense for a farmer on a couple of thousand acres to sell their entire harvest of noodle wheat to one buyer, but that's not necessarily the case for farmers on smaller holdings who need to maximise their income from a smaller amount of produce. Getting into retail doesn't mean you need to become the Coles of cumquats, the Westfield of Wyandottes or the Aldi of achachas; it just means you choose to sell some or all of your produce to members of the public rather than through wholesale markets or other outlets. There are all sorts of ways to do this, and that's what this chapter is all about.

Customers you know

The low hanging fruit in terms of customers are neighbours, family, and the family and friends of people you socialise with through off-farm work, sporting clubs, schools, interest groups and churches. Unfortunately, this low hanging fruit also can put you in the difficult position of selling to people you know, and introduce all the problems that accompany that including pricing your produce too low or giving people more than they have paid for out of a sense of friendship and generosity. Additionally, these customers will only be able to soak up so much produce, and a business targeting these people as your main customers is more likely to be a hobby than a real enterprise, which is likely to be how the tax office will see it too.

Locals

Casting the net wider, you might be able to snare the populations of nearby small towns and passing drivers from outlying farms. Many people living in rural regions already produce their own eggs and lemons, or have an established honey supplier, so you'd need to think about supplying produce that is more difficult to grow or in high demand, or aim to have your produce available a few weeks or a month before the regular season starts, for example, by using polytunnels or greenhouses. You can also reach locals through farmers' markets, farm-gate stalls, CSA schemes and local retail outlets.

Roadside stall

What do you picture when you think of a roadside stall? So many images flood my mind:

- an esky out the front of a gate with a handwritten cardboard sign
- three wheelbarrows loaded with pumpkins
- buckets full of bouquets
- a corrugated iron lean-to with worn timber shelves displaying fresh-picked greens
- a colourful old caravan
- a small three sided tin shed chock full of preserves
- a set of coin-operated lockers
- a vending machine delivering jars.

Roadside stalls come in all shapes and sizes, and are located on the road reserve directly in front of, beside, or just inside the farming property where the produce is produced. Local council, food health and safety, and traffic laws need to be adhered to, as well as the laws of common sense.

Firstly, you need to decide where the safest place to locate the roadside stall is, keeping in mind the traffic flow, clear line of sight, safety for vehicles pulling in and out, and other potential obstructions. You also need to consider signage. Signs cannot be fixed to power poles, road signs or existing trees, but can be attached to your letterbox, fence or gate. The signage needs to be small, and can't pose any risk to the public, for example, by being blown off its moorings. Have a chat with your local council before you get too far into your planning to see if your idea has a chance of approval without a Development Application (DA). Other issues to consider include the following.

- **Weatherproofing**. What is the best way to protect your produce from wind, rain and a searing sun? Perhaps an old fridge or esky under the shade of a tree will do the trick.
- **Rodent-proofing**. How do you keep birds, insects and mice from a free feed, and keep your produce free of contamination? Perhaps some mesh screens and repurposed cupboards will do the job.
- **Theft-proofing**. It's heartbreaking to even think that a person would take advantage of your hard work, but it does happen. It might be the taking of more produce than the customer paid for, or it might be the stealing of cash left by another customer. To mitigate these risks, think about securely attaching a locked receptacle for cash that does not allow access to anything larger than coins and folded up notes (i.e. no hands can get in). Have a sign saying "We clear the cash every few hours and take it to the bank", although some people put up less friendly signs such as "You take-a-my-cash, you take-a-my-bullet". Yikes! I'm not sure how many return customers they get with that type of customer service attitude, but it would make you think twice about stealing! You also need to think through the security issues around people knowing you have cash on-site. This is more of a problem if you become wildly successful than if you are just selling a dozen eggs a day. There are also remote and sensor cameras that can be useful, or you could plan to be at your stall and only open it for an hour a day or at times that suit you.

- **Maintenance.** Add into your schedule the regular mowing and clearing you will need to do to keep access to the stall clear and safe. Customers don't want to be greeted by snakes, clutter and spider webs.
- **Marketing.** Use social media such as Facebook and Instagram to let your local customers know when you are putting out fresh produce. This is a great way to encourage purchases and to help your loyal customers avoid the disappointment of driving to your stall only to find there is nothing there. Hopefully, travellers will see your road signage too and be drawn to stop at your stall, so have another sign there letting new customers know where else your produce is available, for example, farmers' markets or retail outlets.
- **Legislation.** You will need the appropriate Food Authority approval to operate, and will need to adhere to current regulations to ensure that any food products are stored safely. This might mean storing at least 750 mm off the ground and providing protection from contamination and the elements. If you are not going to be talking with people about what they are buying, you will also need to list any ingredients so people with allergies know what is contained in your products. Labels need to include the address where the food was made and a 'best before' date.

Farmgate shop

The next step up from a roadside stall is a dedicated farmgate shop. This is a shop that is within your property boundary and is often a more permanent set up that is staffed. It might be part of an existing shed, or a converted shipping container, or a purpose-built building.

A farmgate shop can operate on a seasonal basis or at certain times on specific days of the week that suit both you and your customers. The best farmgate shops have consistent hours, so customers who make the journey don't suddenly find you closed when they get there. You might find that Saturdays from 10 am to 4 pm work for you, or Wednesdays, Fridays and Sundays from 10 am to 12 noon, or if you're in a well-travelled area, or are always on the farm and can be doing other jobs while the shop is open, you might run a seven-day-a-week operation. Some type of buzzer or camera system can alert you to when customers enter the farm so that you or your staff can stop what you are doing and serve in the shop. Or, if you are successful enough, you'll always have someone at work in the shop.

A farmgate shop gives you the opportunity to meet and interact with your customers. This can be a great joy, and can give you insight into the products customers want. Alternatively, it can be a huge drain on your time if a customer spends a few dollars and hangs around for ages asking you questions. You might find it frustrating to be caught up in a leisurely chat when you should be out on the farm furiously planting so you have something to sell in a few months' time. And just as that customer finally leaves, another one of similar garrulity takes his place!

A benefit to having your own on-site shop is that you won't need to spend time travelling to farmers' markets, paying stall fees and being away from the farm. That said, farmers' markets offer plenty of opportunities for success because of the large amount of foot traffic they attract. It's a lot harder to get the same number of customers out to your farm. To do so, you need to be media savvy, join in publicity efforts with tourism bodies, potentially welcome coach tours, and also be an active participant in local food-trail promotions.

When you market your farmgate shop well and word travels, you may end up receiving visitors at unexpected times. They will say things like, "we noticed you were closed, but hope you don't mind, we just wanted to buy...". Or, "sorry, we let ourselves through the gate, we just love your...". Or they might knock on your front door while you're sitting down to dinner on a Saturday night or while still in your pyjamas on a Sunday morning. You can minimise these intrusions by padlocking gates or having a sign saying something like "if the gate is closed and you come in, expect to be composted". Not the friendliest message, but sometimes you need to be firm to maintain your privacy and sanity.

If you are planning a farmgate shop from scratch, think about how and where you can site it most efficiently on the farm, and how it can support your workflow rather than detracting from it. For example, a cheesemaker might have Perspex windows looking out to the farm shop so people can see cheeses being made without interrupting the workers.

Likewise, workers can see when people enter the shop, and can come out to serve them as needed.

The safety of customers, farm biosecurity and general security all need to be considered when inviting customers onto your land. Little things like strategically locating the cash register so you can see what is going on is a good idea, but you also don't want to place it in a position whereby customers purchasing products block the way of new arrivals wanting to enter the shop.

You will also need insurance for public and product liability.

Often, it is only after you open your venture that you will understand the impact of having customers coming onto your farm. You might also discover other opportunities such as offering farm tours for a fee.

Community Suported Agriculture (CSA)

The CSA farming model has captured the attention of trailblazers, activists and passionate foodies everywhere. In 1986, there were two known CSA's in the United States, and now there are more than 12 000. The numbers are nowhere near that in Australia, and a number of those that were first written about in 2014 are no longer operating, but the concept has caught on.

The idea is that customers pay up front to help a farmer get through the season (knowing full well that there are risks due to climate, pests and other woes), and in return the farmer provides a box of produce each week. This varies from CSA to CSA, as some accept weekly or monthly payments, while others also ask members to pitch in and help for a few hours on the farm. By receiving money up front, the farmer can plan the season and concentrate on servicing a loyal customer base rather than chasing customers and markets.

What do customers love about CSA's? They enjoy the connection of knowing how, where and by whom their food was grown. They also delight in experiencing the ultimate in fresh, local nourishment.

What don't customers like about CSA's? Not knowing exactly what will be in the box next week...and that because it's purely seasonal produce, some 'staples' won't be included.

Many of the CSA's in the US only operate during spring and summer, whereas in Australia they are more likely to operate through spring, summer and autumn, with the farmer taking a break in winter. CSA's work best near larger population centres and regional towns.

You can have a vegetable CSA or a meat-only CSA. It's also possible to operate a multi-farm CSA whereby meat, veggies, eggs, fruit and even flowers and fibre are produced in cooperation with other local farmers.

Marketing is mostly done by word of mouth and local publicity. What follows is a summary of some of the complexities of CSA's.

- One acre can support between 20–35 households.
- You need to balance growing what people want to eat with growing things that are easier for you to produce. You won't be able to provide everything, e.g. all fruits, enough potatoes etc, but by clearly explaining what you can produce, customers won't be let down.
- Volunteer and paid internships can help with labour costs.
- Inform members about the vagaries of farming and what might be available in the boxes.
- Provide recipes (especially for your more unusual produce or when there is a glut of something), newsletters and open days to develop customer loyalty and knowledge.
- Strike a balance between forming a connection with customers and people thinking they can just drop in at any time.
- In addition to labour costs, you also need to plan for items such as a walk-in cooler, irrigation, greenhouses, storage, a website/newsletters, and your time managing the CSA. You'll also need to decide whether you will deliver the boxes of produce or customers will pick them up at a set time.
- Can you mentally handle 'owing' customers something after they have paid up front?
- Do you have a strong sense of mission/purpose/values?
- Can you handle finicky customers, and do you have the patience to communicate with them?

Here are five CSA tips from Peter Carlyon from Transition Farm on the Mornington Peninsula in Victoria.

- Know that you can grow the food.
- Start on a small scale.
- Educate CSA members about what they are signing up to.
- Try to diversify the boxes as much as you can.
- Love hard work!

CSA's are a great way to build a strong connection between your farm and your customers, but to run a successful one you need to be extraordinarily hardworking, positive and a great planner.

FARMER SPOTLIGHT: On-farm butchery and CSA

A cultural studies academic and former vegetarian gleans wisdom from other farmers, apprentices to a butcher, runs a thriving meat-based CSA and uses crowdfunding to establish an on-farm butchery.

Farmer: Tammi Jonas
Farming: Heritage Large Black pigs, cattle and chickens
Farm name: Jonai Farms & Meatsmiths
Farm size: 28 ha (69 acres)
Average annual rainfall: 960 mm
Where: Eganstown, Central Highlands, Victoria
Farming since: 2011

Volcanic soil, combined with ethical and ecological philosophies, underpins the success of Jonai Farms. With no off-farm income, the Jonas family (who call themselves the Jonai) make a living from their land through a combination of farmgate sales, sales to chefs and via a thriving CSA (which is so popular it has a multi-year waiting list). They also offer Grow Your Ethics producers' workshops at the farm to teach their entire model.

Raised on a cattle ranch in the US, after spending two decades in the city Tammi grew to loathe industrialised intensive animal agriculture. She and husband Stuart were galvanised by a Joel Salatin talk to raise free-range pigs (they have 12 sows, two boars and 100–120 growers at any one time), and then started in cattle to manage the grass.

"The cattle became an important offering to diversify our produce, and we added layer chickens in 2015 as an ecological service to increase the fertility of our paddocks," said Tammi.

In 2013, the Jonai successfully crowdfunded major infrastructure for their farm, raising $27 570 on the website Pozible from 166 supporters (many of whom became loyal customers). The funds were used to convert a 40-foot refrigerated shipping container into a butcher's shop.

The Jonai are continually innovating. They have moved from purpose-grown grain for animal feed to so-called waste-stream feed that would otherwise end up as landfill. "By working with the farming and business communities, we tap into landfill-bound food such as spent brewers' grain, post-harvest surplus fruit and veg, and supply-chain-damaged dairy produce, even water-damaged rice," said Tammi.

Tammi works up to 50 hours per week (50% physical labour and 50% on the business). A typical week might involve butchering pork and beef, making sausages, bacon, ham, smallgoods, charcuterie, bone broth and pâté de tête, delivering mixed cuts to hubs and communicating with CSA members. There's also farm planning, accounts and invoicing, updating the website, writing blog posts, social media updates and advocating for food sovereignty to be done. Tammi particularly enjoys sausage recipe development, noting what's coming into season and developing recipes to include it in their sausages, such as locally grown and made pomegranate molasses, and chestnuts, apple and sage.

It's a busy life, but they get time away thanks to Stuart's supportive parents, staff, interns and the local community. "I love everything about farming, but it's still essential to our family's emotional and physical wellbeing to take time away together and renew and gain fresh perspectives," said Tammy.

Food for thought from Tammi

What are the basics required for your farming activity?
Physical strength, as well as knowledge of soils, pasture grasses, building, plumbing and animal health and nutrition. In our system, you also need butchering, cooking, teaching and compliance skills. To help you get

started, the Deep Winter Agrarians group on Facebook is an amazing resource for the small-scale farming community. You can also refer to old farming books from the first half of the 20th century, as their animal husbandry advice isn't about intensive piggeries.

What are the threats to the industry?
The biggest threat to small-scale farming is inappropriate-to-scale regulation. This hinders our efforts to grow fair food and sell it directly to eaters.

Opportunities in the industry?
There is a growing awareness of the problems of industrial food systems. People want to know more about where and how their food is grown, and by whom. The CSA model provides these answers. You can also take control of your value chain, as we have, by building a boning room and commercial kitchen.

What makes a good farmer?
Someone who's open and curious, and who values the satisfaction that comes from hard work. It's also about being thorough in everything from how you attend to your animals or crops to how you keep and analyse your finances. A good farmer is someone who puts the welfare of soils, waterways and animals first.

Best pieces of advice from another farmer?
Don't pay for advertising. Don't make farmers' markets your primary outlet. Pigs do well on a pastured diet with milk and eggs.

Worst piece of advice from another farmer?
All advice is useful, as it makes you think about the alternatives.

What would you farm if you weren't involved in your current activity?
We want a diverse, agroecological farm, so the only thing we would like to do is increase the diversity of what we farm. We're trialling some garlic and planting loads of fruit and nut trees for animal fodder. I'm also working on a book called Fair Food Farms.

Tammi's top 5 tips

1. Focus on the parts of your system that will make you profitable first –
you have the rest of your farming life to improve your ecological services
that don't have a clear return on investment (ROI).

2. If you're going into livestock, choose animals you like, and don't
attempt too many species at once – introduce each one slowly to best
manage the labour and input requirements such as fencing, watering,
housing, feed, moving animals etc.

3. Formulate a reasonable model for the size of operation you intend to
achieve and monitor your growth in tandem with monitoring the impact
of your agricultural activities on the environment. Avoid a growth
mentality.

4. Keep good records – financial, livestock, feed, risk management etc.

5. Consider developing a CSA model once you've established your
supply of produce. It's a risk-sharing model, with the added bonuses of
building community, educating people, and reducing the administrative
load.

What's the best way to enjoy your product?

Pulled pork tacos. Who doesn't drool just reading the words 'pulled pork
tacos'? I make them whenever I have bits of pork to use up, such as some
forequarter chops with freezer burn. The beauty of pulled pork is that
you can make it with any cut or mix of cuts, although shoulder is best,
and you can cook it dry or wet. The possibilities are endless! It's great for
barbecues or Mexican-style, as sliders or tacos! Any leftovers are always
beautiful in enchiladas!

The trick with pulled pork is to cook it low and slow, whether wet or dry,
until it's so tender it pulls easily with a couple of forks. One common
version I do is to take a banjo (shoulder joint) covered in chimichurri and
salt, wrap it in foil and slow cook it for no less than five hours. Tender
goodness! The method below is the simplest I know. Chuck everything
in a big cast-iron enamel pot, pour a bottle of cider over it, pop it in a low
oven (140 C) and a few hours later, you've got an amazing lazy delight!

Ingredients

1 kg of ethical pork, bone in (shoulder is best, but any cut will do)
1 bottle of cider (beer works fine too)

2–3 tsp oregano
bunch of parsley
2 onions, roughly chopped
5–6 cloves garlic, lightly crushed
1–2 chillies, chopped
splash of cider vinegar
salt to taste

Method
If it dries out during cooking, you can add more cider or beer after the
first couple of hours.
When meltingly tender, pull apart with two forks, discarding the bones.
Season to taste.
A classic combo for these tacos is simply pulled pork, shredded cabbage,
tomato, red onion, salsa verde (from tomatillos) and guacamole, but we
use whatever greens we have around, and we're fans of homemade
yoghurt instead of sour cream. A bit of Tabasco and you're in heaven!

~

Crowdfunding

Crowdfunding is both a way to access funds and a brilliant loyalty
marketing tool...but first you must have an idea, a product that captures
the imagination of your supporters and plenty of admin time!

The ATO defines crowdfunding as "the practice of using internet
platforms, mail-order subscriptions, benefit events and other methods to
find supporters and raise funds for a project or venture. If you earn or
receive any money through crowdfunding, some or all of it may be
assessable (taxable) income, depending on the nature of the arrangement, your role in it and your circumstances. All assessable income
needs to be declared on your tax return. Similarly, if any amount is
assessable income then some of the costs related to gaining or producing
that income may be allowable deductions, providing you have the appropriate records to substantiate your claims."

The steps to a successful crowdfunding campaign include the
following.

Customers. Build a base of customers who like what you do and your approach to farming. Ensure you can contact them through a combination of email lists, your website, social media and news media. You will need these people to be your initial backers and to spread the word, otherwise it's going to be very difficult for your campaign to rise above the thousands of other hopefuls.

Come up with a big idea! What is it you want to deliver? What infrastructure does your farm need to do it? Be reasonable and rational because if you don't reach your funding target, you won't get any of the funds. Think about what will inspire your customers to pledge their money, don't just think about your own needs.

Think it through. Think deeply about what will entice your backers to fund you at the different levels of sponsorship. It needs to be a win-win, e.g. a backer of $10 will receive a handwritten thank you postcard (with a picture of your farm) and a mention on your website and social media. A backer of $50 might receive the above as well as two of your products. A backer of $250 will receive all the above plus extra products. A backer of $1 000 will receive all the above plus a private tour of the farm, plus a hosted lunch with local wine and a driver to take him or her back to their accommodation. There are so many different options. What are some unique rewards you can offer backers that can't normally be bought? Perhaps you will name an animal after them, put their name on a wall, plant a tree on their behalf, send them a bottle of your best tipple with their name in the label. Work out what will motivate people to support you. You can also set limits on how many rewards you have at each level.

Make a plan. Intricately plan your project (including a budget incorporating the cost of fulfilling your promises to backers) because in addition to your time, there will be costs incurred including the fees you will need to pay the crowdfunding platform.

Research. Scope out the different crowdfunding platforms (such as Pozible, Indiegogo, Kickstarter, GoFundMe) to determine the best one for your campaign.

Share your vision. Bring you vision to life with emotionally resonating text and video.

Promote. You then need to promote the hell out of it through social and news media. The project can be small or large. It just needs to capture the imagination of supporters and reward them in some way, either by making them feel like they've contributed to a great cause or physically rewarding them with products and experiences.

Communicate. Communication is key, and factoring in any potential delays when setting your estimated delivery date is important to retain customer satisfaction. Dealing with lots of backers means you need to be on the front foot, otherwise you'll have an administrative and PR nightmare on your hands. Deliver what you said you would, and then go beyond your backers' expectations, and you will have customers for life.

Fulfil. Just to make sure you got the message, ensure that you can deliver the goods, both literally and figuratively! You will need to deliver what you promised or refund your backers and suffer negative consequences.

Kickstarter says that launching a Kickstarter crowdfunding campaign is a very public act, and creators put their reputations on the line when they do so. "Backers should look for creators who share a clear plan for how their project will be completed and who have a history of bringing their creative ventures and other projects to fruition. Creators are encouraged to share links and as much background information as possible so backers can make informed decisions about the projects they will support. If a creator has no demonstrable experience in doing something similar or doesn't share key information, backers should take that into consideration. Does the creator include links to any websites that show work related to the project, or past projects? Does the creator appear in the video? Have they connected via Facebook?"

You can research examples of successful agricultural campaigns from around the world on the internet, but the following information is an example of a small farm that was obviously well-embedded and highly thought of in the community and has run two successful Kickstarter projects.

Project name: Grade A Garden Eggs Project (2015)

- Aim. To double the number of birds in their laying flock including building a mobile coop and purchasing fences, nesting boxes and feed to raise the chicks to laying age over winter.
- Region. Des Moines, Iowa
- $8 887 pledged towards the goal of $8 500 by 63 backers.

Pledge levels offered were:

- $20+ Name a chicken, receive a certificate with your bird's face and name.
- $50+ Receive an organic cotton Grade A Gardens t-shirt.
- $125+ Receive a pound of gourmet seed garlic and get a personal tour of the farm and tips on how to plant your garlic so you can grow it at home.
- $250+ Receive a t-shirt plus 49 lbs of vegetables including garlic, shallots, onions, potatoes, sweet potatoes and winter squash.
- $500+ Receive a t-shirt plus 49 lbs of veg PLUS an Autumn CSA share (four weeks of fresh veg).
- $1 000+ Receive a t-shirt plus a full-season CSA share plus one dozen eggs a week for the entire 2016 season.

Project name: Grade A Gardens Greenhouse Project (2017)

- Aim: to build a greenhouse for certified organic vegetable production and to host farm dinners by local chefs.
- $18 000+ pledged towards the goal of $15 250 by 70+ backers.

Pledge levels offered were:

- $20+ a cute three-bulb garlic braid to hang in your kitchen.

- $50+ a box of storage produce (onions, garlic, potatoes etc.).
- $100+ a box of storage produce and a beautiful 12-bulb garlic braid with flowers.
- $200+ a seat at the farm's inaugural Greenhouse Dinner.
- $200+ stock your own garden with plants available from the farm's plant sale.
- $300+ dinner for two at the inaugural Greenhouse Dinner.
- $500+ start a garden: help with tilling, planning and planting your garden including plants and two extra consults and weeding help during the growing season.
- $1 000+ a full-season CSA share + an autumn CSA share + one dozen eggs a week + one seat at the inaugural Greenhouse Dinner.
- $5 000+ a lifetime CSA share, every week, every year + one dozen eggs a week + two seats at the inaugural Greenhouse Dinner.

From artisan cheesemaking kitchens to food trucks, from on-farm butcheries to farmgate shops, from the purchase of stud animals to the installation of solar-powered irrigation, you might be able to give your venture a kick start with crowdfunding.

Online retail

Do you have easy access to Australia Post or a courier company to help ship your produce? If so, online retail might be for you. Even though it's easy to set up your website for ecommerce, it can be time-consuming in the initial stages. Here are some things you need to spend time on to get your online retail offering right.

- Carefully weigh and measure items to correctly work out postage or courier rates; that way, you won't undercharge the customer and short-change yourself.
- Factor the cost of packaging your goods, in terms of both labour and materials, into your online prices.
- Take great pictures of your product against a clear white background. Show it in the ground, how it looks packaged

and how it is used. Great photography will help you to sell more product, so don't skimp on your photos.

- Showcase customer testimonials.

In addition to your retail efforts on your own website, consider working with suitable aggregators. At the time of writing, Australia Post's Farmhouse Direct is doing a great job of connecting a huge customer base with farmers across the country. For a reasonable commission on sales, you can be included in their marketing efforts and reach a whole new audience. In addition to Farmhouse Direct, there are many other sites on the internet that you can sell through, so research their charges, commissions and ease of use before signing up. One to take a good look at is the ecommerce solution provided by the Open Food Network.

Pop-up shop

Pop-up shops aren't just the domain of hot fashion labels. If you have a bumper harvest, why not rent a small space or a table at your local shopping mall, or ask the local council or real estate agents if there are any unleased spaces you can have on a short-term rental. How about a marquee in a church carpark or at a local sporting event? Pop-up shops, when done well, enable you to generate publicity (via a media release about your venture), create excitement and access great foot traffic. It's a way of promoting your farm and produce in a fun, creative way. Join with others to make a real impact!

There's no use growing produce if you can't sell it. Another way people reach customers directly is through farmers' markets, and given the rising popularity of markets, the next chapter is devoted to these as an outlet.

21

FARMERS' MARKETS

The interest in food and farming is hot right now, and that in farmers' markets, where they are brought together, is even hotter. I'm writing this chapter at a farmers' market while my 15-year-old daughter runs the stand and, I swear, it is hot, really hot. It's 38 C in the shade and even though the canvas roof is blocking out the sun, nothing can block the heat radiating up from the concrete pavers. It helps on days like these to remember the bone-numbingly frigid markets we've also attended where frost clung to our boots for hours, when winds blew tents into rivers and when not even gumboots and double raincoats could keep us dry from downpours. That's the thing about farmers' markets; they can be brutal or bountiful, and sometimes – actually, often – both.

We've driven the ute to, set up for, and worked at more than 600 farmers' markets. We've experienced small rural markets like those at Gloucester and Nabiac through to large city markets such as The Beaches Market in Sydney and the Newcastle City Farmers' Market. 600+ markets doesn't make us veterans, but it has given us lots of insights into how farmers' markets can both buoy and/or sink a farm strategy.

Something we observed early on is that fresh and cooked produce sells, but so can other items ranging from iron-sculpted whale tails for the garden to flowers and compost worms. Trends come and go; one year it might be kefir, the next kombucha, the next cold-pressed coffee, but what always stays the same is that customers are looking for great produce.

Markets come in all shapes and sizes. Some are held in parks, by the water, at sporting ovals, at showgrounds, in halls, under pavilions, and at racecourses, wineries and schools. Some market managers specify a certain type and colour of tent, while others let you bring a standard-size tent (normally 3m x 3m) in any colour you like. At some markets, you need to bring your own floor covering, while others don't mind if you don't bother.

It took us many markets to work out why we weren't making any money, one reason: profitless volume! That's called slow learning, and therefore our intent in this chapter is to accelerate yours.

Why markets?

Some of the key benefits of including a market in your farm's business strategy include the following.

- Markets offer access to vast numbers of customers, which is especially important if your farm is off the beaten track.
- You can sell direct to the consumer, without any middlemen.
- You can test products in small batches.
- Direct contact and feedback from customers help you sort out recipes, packaging and pricing.
- You can move your seasonal harvest.
- You can provide employment opportunities for other family members.
- A market gives you the ability to promote and market your business while earning income.
- Camaraderie with other stallholders can buoy you.
- Cooperation and networking with other stallholders can lead to business opportunities.
- At a farmers' market, you're right at the coalface. You can see trends develop and dwindle, you can see what sells and why, and you can learn from other stallholders.

- You can buy produce at mates' rates from other farmers or trade items.
- Markets give you cash flow.
- A good market stall gives you visibility with retail buyers, chefs and exporters. A market is an opportunity for your product to be discovered.
- You're able to connect directly with customers and build amazing loyalty (be sure to get their email).
- You become a core member of the community.
- It can be fun and uplifting to get off the farm and meet happy customers.
- You might be able to share a stall with another farmer from your region so you can alternate markets, or hire someone by the hour to take some of the load off you while still reaping the benefits.
- Markets force you to focus on your product, marketing, financial and sales skills.

Why not markets?

Unfortunately, there are probably just as many reasons not to do markets as there are to do them! Here are some of them.

- Long travel time to markets leads to wear and tear on both vehicles AND you.
- Attending markets means you are away from your farm, which can be risky if your animals are due to give birth or if there are fires or floods.
- Competitors can quickly see what you are doing and try to replicate it.
- It can be heavy physical work packing and lifting boxes.
- You need to have a great day to make the hours you spend packing the truck or ute, restocking boxes, driving and staffing the stall worthwhile.
- Markets are often held on weekends, which means you may need to compensate staff at a higher rate. If you work during the week too, it means you are working seven days a week.

- Inclement weather can make or break your sales. If it's rainy in the morning, or boiling hot, people often won't come out. Meanwhile, your products may get damaged in either case.
- Some markets allow a large number of stallholders to compete with the same products, which means there is less profit to go around.
- Shoplifting happens at markets too, so you can lose not just your produce but your trust in people.
- Some customers can be incredibly rude and irritating, and when you are standing behind your stall for hours on end it can seem like there is no escape. This is where humour helps!
- After a long day, you still have to get back to the farm and unpack and get on top of your regular farm chores.
- Some farmers' markets attract very price-sensitive customers, which means your margins might be squeezed.
- A run-in with an overzealous market manager can see you banned from the market or relegated to a site in the dusty back corner behind the bins. Market management can also change, and they can bring in new competitors, making things tougher for you.

Sometimes, there is a clear barrier to entry in that there is no space in the market or too much competition in your field and you are not allowed to participate. In that case, you either need to be patient, approach a different market, or even think about setting up your own market in a good location. However, this is a last resort, as market management takes a HUGE amount of time and great people management skills.

Which market?

When choosing a market, you might be limited to the one closest to your farm, as it's the only one you can find the time to do, or you might be able to look further afield and choose a variety of markets to attend. The further the market is from your farm, the more sense it makes to tag additional markets into the same trip so your travel costs are split between two or even three markets across the weekend.

It makes sense to visit a few markets across a region to get a feel for:

- The main customer type. Are they foodies or bargain hunters? Tourists or young families? City singles in yoga pants or country dudes in their boots? Are there organic evangelists, paleos or multicultural buyers?
- The "vibe" of the market. Are you nearly run over by bustling, business-like shoppers, or are they relaxed and chatty? Is the market loud and crowded, or quirky, casual and fun?
- The frequency. Is the market monthly or weekly? Are you able to attend just one market a month or do you need to attend each week? What would work best for your produce and schedule?
- The cost to attend. Although a city market might double or quadruple your sales, by the time you add in your petrol, driving time and the higher cost to attend the market, is it really worthwhile?
- The kinds of displays stallholders are using. Do they pay attention to decoration and signage or is it more about bulk produce? How might your stall look? Which stalls attract you with their layout?
- The price points for produce similar to yours. Often, the produce at country markets is priced a lot lower than city markets can sustain.
- How many customers are coming through the gates, and what times seem busiest? I found that at one city market I attended I'd be run off my feet between 7 am and 9 am, while at our local village market people weren't really getting going until 9.30 am...make that 10 am on a cold winter's day!
- How established the market is. Is it a market that has been operating for a long time or is it a newer market? Sometimes it can be hard to gain traction as a new stallholder at a long-running market. That's because so much loyalty has already been built up by the regular stallholders and customers make a beeline for the same produce every week, practically wearing blinkers in relation to any other stall. You can break through, but it will take extra work on your part. If, however,

you are at a newer market, you can begin to develop loyalty from the start, but it can take a long time for these markets to reach the same level of customer traffic.

- The market operators. How do the market managers seem to treat stallholders and customers?

Stall look and feel

In any shopping centre, Boost Juice is instantly distinguishable from Muffin Break, which in turn is instantly distinguishable from The Coffee Club. One look in a clothing retailer's window will tell you whether they are targeting retired women, teenage girls or sporty types. A browse along the shelves in a supermarket shows packaging aimed at budget through to luxury shoppers. How stores dress their windows, what music they pipe, the coverings on the floor and walls through to how products are stacked – these are all elements of merchandising.

Big businesses invest heavily in brand and visual merchandising to build strong identities to attract their target market and build loyalty. They want to cut through visually and entice purchasers into their store. Likewise, at a market, your stall needs to instantly cut through the clutter and be distinguishable. It needs to project your unique image in a way that appeals to and resonates with the customers you are there to tempt and convert. And you thought just growing the produce was a challenge!

There are many elements to getting the look and feel of your stall right, but the main areas to focus on are:

- being true to your brand
- merchandising: product display and overall look of the stall
- protection from the elements
- customer flow and ease of shopping experience
- provision of a safe workplace.

You can create a great farmers' market stall using the following guidelines.

The stall itself

- A fully screen-printed marquee.
- A vintage circus-like tent.
- A custom-made or repurposed caravan, Kombi van or other vehicle or trailer.
- A decorated truck.

Colour

- Use a colour that mimics or accentuates what you are selling. Selling strawberries? Go for red to own the whole strawberry theme, or black or white to show off the red even more. Selling citrus? Use orange or white.
- Natural tones evoke earthiness. Moss to forest greens, earthy creams and browns enhance the message of naturally produced and organic produce.
- Tangy limes, bright yellow, hot pinks and other lively colours can stand out.
- Red-and-white checks make people think of country and home-style cooking.
- Purple oozes chocolate.
- Black and white speaks to professionalism, a high level of service and clinical attention to detail.
- A little tip to remember is that lighting is important in a stall, so choose a light-coloured marquee roof so that you still get some natural light into your stall on overcast days.

Fabric

- Table coverings need to go all the way to the ground on the customer side to hide the clutter. Make sure it's not so long that people can get their feet tangled in it (sometimes this requires tying up the corners if your table covering isn't custom-made).
- Hessian is a favourite of many an organic grower, but this can also make it hard to stand out unless you have other compelling features on your stall.

- Old sacks evoke rustic and natural, and your commitment to re-use.
- Velvet suggests decadence.
- Synthetic, stain-resistant tablecloths are easy to clean, but other stallholders might have purchased the same cloths. They come in a variety of colours though, which is handy.
- A patchwork of fabric might be interesting, but like any heavy pattern, it can look visually cluttered and detract from your produce.
- Layers of fabric allow you to create different effects.
- A fully screen-printed table covering allows you to showcase images of your farm or your produce or use large lettering so people know exactly what you are selling.

Images

- Images of your farm and produce will really resonate with customers and help tell your story. They can also be calming and motivating for you to look at when you're a long way from home on a quiet day!
- Connect your produce with its place of origin by positioning photos of your farm around your stall. You can do this through individual photos or by having them printed and turned into large banners and professionally printed tablecloths.
- Some stallholders run looping audiovisual presentations on iPads and tablets.

The senses

- Appeal to the senses. How can you use sight, smell, sound, taste and touch to connect with customers?
- Offer samples to enliven taste buds.
- Create captivating smells by regularly cooking samples.
- Add drops of essential oils around your stall or crush herbs regularly.

- Burn candles (be careful of fire and hazard risks from nearby gas bottles etc.).
- Play music or make music (not overly loud or in a way that will drown out speech or annoy your neighbours).
- Add a wind chime.
- Have moving whirly birds or fluttering flags above your stall to draw attention from further up the aisle.

Unique receptacles

- Baskets and crates are great, as they are safe and easy to select from.
- If you don't have enough produce to fill a basket, fill it with straw or bark and place the produce in the inner ring – this way, the baskets will still look full and inviting.
- Try out French provincial-type baskets or handwoven baskets.
- WWII ammunition crates and timber crates you can make from pallets or fence palings can look great, as can rough-hewn logs.
- Metal cages can be made from fencing wire.
- Clear plastic/Perspex cylinders or stackable containers can display your produce in an interesting, dust-free way.
- Large recycled catering tins, cut-down wine barrels, old saucepans and camp ovens can all be used innovatively to display produce. We've even used something as simple and abundant as egg cartons to show off our lip balms and tea lights.

Signage

- A key thing about stall signage is that it's less important to highlight your company name than it is to focus on WHAT you are selling and your farming approach. As an example, if your farm name is Harry's Orchard, you might have that in small writing, but in large writing you would have HERITAGE APPLES, APPLE PIES, APPLE SAUCE, and DRIED APPLE.

- You won't be able to talk to every customer, not just because you might be busy serving someone else, but also because they might not give you the opportunity! For every customer who wants to ask you questions and gets a chance to, there is a customer who will browse from a distance. This is where great signage becomes even more important.
- Explain through signs why your produce looks the way it does, e.g. our strawberries are small because they are a heritage variety that tastes BIG! They are not pumped full of water and are delicious in salads, with custard or just on their own.
- Make your signs easy to read so people don't need to squint. Large, easy-to-read lettering is best.
- You can handwrite them on cardboard, paint them onto anything solid, print them in a funky font, chalk them onto boards, or laser or screenprint them onto banners.
- If using chalkboard signs, regularly clean and repaint the black so that the lettering stands out.
- Hand-painted signs, if done well, can add more character than professionally printed signs.
- Think about waterproofing your paper/cardboard-based signs by laminating them so they are easy to wipe clean and don't get damaged by rain.
- People don't want to have to ask about prices; they want to be able to see them. So make sure that the prices of all your products are clearly marked.

Stacking

- You need to create a great display to create interest.
- "Stack it high and watch it fly" is a retail saying that speaks to the appeal of abundance. People will be attracted to your stall by its look of fullness and generosity, not its scarcity. If you are selling something small like wattle seeds rather than watermelons, create a display that showcases them.
- Stack carefully so that it's easy for customers to pick up items without worrying that the whole display will collapse.

- Refill regularly so you maintain the look of abundance.
- Head outside your stall and look back at it from the customer's viewpoint – it's amazing what you'll see that needs fixing. Make changes such as turning labels to face customers.
- Towards the end of the market, transfer items from big baskets to smaller baskets so they don't look like the leftovers nobody wanted. Re-stack items and cluster them together.
- Although a look of abundance is the overall aim, less can sometimes be more. Too much clutter looks unprofessional, and it can be hard for people to get a handle on what it is you're selling. If you have a lot of items, cluster them by colour or shape so that your stall looks more ordered, striking and engaging to the eye.

Height and depth

- Go up and go out or go home! The use of "risers" helps to create height and depth for your display.
- "Risers" are basically upturned boxes or buckets, stands and shelves that create different levels and depths at your stall.
- Think about where people's gaze naturally falls...it's not down on the table, it's pretty much at eye level.
- If you can, angle your baskets and crates so they aren't flat on the table but are slanted up towards the customer's eyes.
- It's rare for customers to want to look directly downwards, so come up with ways for them to be able to look forward such as putting racks or a small bookcase or an open cupboard on top of your table. Keep it to the side though so people can still engage directly with you.
- Think about how you will transport your produce, as the way you bring your stock to market, for example in boxes, crates and buckets, can become the basis of your risers and overall display.
- If your risers are plastic crates, once you unload them and turn them over, cover them with a fabric that enhances your stall.

You and your staff

- You are on display too!
- Take pride in your product and show respect for it by carefully placing items in their display receptacles or in customer bags.
- Wear a uniform, a cap or something that indicates that you are running the stall, and are not just another customer.
- Ensure that your clothes are clean and hygienic because although customers love the idea of farms, they don't like the idea of eating muck.
- Make sure your staff know everything there is to know about the produce on your stand.
- Ensure that you're comfortable. This means wearing thermals on wintry days and light clothing on hot ones.
- Sturdy, enclosed shoes are necessary for comfort and safety.
- Customers want your produce, but they also want knowledge and a smile.
- Markets can mean long days, and you will need to sit down at some stage, but when choosing a seat, pick a high one like a bar stool so you look more engaged. Customers might not want to disturb you if you look too relaxed slumped down there in your deckchair! It's also easier to get off a higher chair than a lower one.

FARMER SPOTLIGHT: Certified Organic Macadamias

From living rough in a tent while he established his farm to winning Champion at the Royal Hobart Fine Food Show for his value added macadamias, Dave Flinter's farming business continues to grow.

Farmer: Dave Flinter
Farming: Macadamias, garlic
Farm name: Hand n Hoe
Farm size: 600 acres (40 acres under planting)
Average annual rainfall: 1 818 mm

Where: Comboyne Plateau, Mid North Coast , NSW
Farming since: 1979

Dave Flinter's farm lies in a relatively isolated, mountainous region just inland from the NSW coast. On approach, everything about the farm looks raw, natural and overgrown, but this organic certified farm runs on permaculture principles and it's the way it's meant to be.

"Vegetation and weeds are all utilised as organic matter eventually," said Dave, "and the best thing about being isolated is we don't need to deal with run off or spray drift from other farms. That's crucial for organic certification."

The farm also benefits from high rainfall and pristine water from two spring-fed creeks.

"Water is gold," said Dave, who delights in the abundance of yabbies, frogs and fish. He also revels in a close relationship with the land, having lived on it in a tent for two years while he cleared small pockets of the rainforest and felled selected timber trees to build his home. That home is now equipped with an impressive solar array and battery back-up system and the macadamia trees he planted grow in amphitheatres throughout different paddocks among the natural vegetation.

Working 60+ hours per week (currently 50% on physical labour and 50% on the business – though it was 90% on the physical side while he was establishing the farm), his weeks entail fertilising and pruning, value adding products such as making nut butter spreads, packaging, packing for and attending markets, preparing orders for retailers, and dealing with the paper trail and bureaucracy. David sells via retail outlets, online and through farmers' markets but has decided not to export due to his focus on local, sustainable food systems.

"Some people might suffer from the culture shock of isolation moving to a property like this," said Dave, "but though it's heavily labour intensive, it's just so rewarding. There are so many health benefits to farming too, and the positives far our weigh the negatives."

Food for thought from Dave

What are the basics needed for farming what you do?
Labour for picking, broad shoulders, patience and passion, as well as

good volcanic soil, high rainfall and the ability to find outlets to sell your produce.

What are the threats to the industry?
There's a lack of government support for organic farmers, little protection and then sudden changes of rules such as the Backpacker Tax. There are also mass plantings of macadamias taking place in other countries around the world.

Opportunities in the industry?
Demand for nuts in general is increasing and there is overwhelming demand from international markets.

What makes a good farmer?
Knowledge, patience and an understanding of the land.

Best piece of advice from another farmer?
Be patient and don't feel bad about making mistakes, trial and error is sometimes what it takes.

Worst piece of advice from another farmer?
We were advised to grow trees that thrived in another area, but discovered they were the worst choice for our particular climate. That taught me to do small trials first to see what the best options were and we now grow six different varieties.

What would you farm if you weren't involved in your current activity?
It wouldn't be something intensive, probably more garlic and sweet potatoes.

Dave's top 5 tips
1. Don't rely on government or councils to support you on your journey, you'll be disappointed if you rely on the system.
2. Trial a lot of different things and diversify.
3. Give different farmers' markets a go, but give them a good crack as you won't see results in the first few months. It takes time to build customer loyalty so be consistent and keep turning up for at least 12 months.

4. Establish a reliable working base with local people as your kids will eventually grow up to go on their own journey, and short-term labour supply has its own challenges.

5. You'll be disappointed if you expect to make a lot of money straight away. Understand that it won't be easy and that you will need to be patient, but if you get it right, you just can't beat the lifestyle.

What's the best way to enjoy your product?
Eat them.

Market managers

Markets can be run by commercial operators, volunteers, council employees or groups of farmers. The market managers are responsible for the success and safety of the market, and they achieve this by scrutinising potential stallholders and ensuring that current stallholders are acting appropriately. They promote the market, organise access and so very much more. It's an incredibly difficult job, and some of the best at it would have to be the team of Bianca, Phil and Lindsay from The Beaches Market. This is a weekly market held on the northern beaches of Sydney at Warriewood. It is an outdoor rain, hail or shine market. Here are their top 22 tips for stallholders.

1. A good display is one where the products are easily visible.
2. Use clear, concise signs.
3. Keep your stall neat and tidy and as natural looking as possible.
4. Polystyrene boxes don't look good and aren't good for the environment.
5. Provide information about what you do and how you do it through signage and discussions – our customers love asking questions of farmers.
6. Be polite and friendly and sell a good product.
7. If you're not a people person, don't have a market stall.
8. Verbally abusing the market manager, customers or other

stallholders will get you banned from a market. So too will continually disobeying simple instructions such as not smoking, not wearing shoes, wearing aprons to the toilet etc.

9. Some of the reasons stallholders fail is that they overinflate their price, they're rude, they diversify too much (e.g. if you are known as the potato man, don't lose focus and start bringing lettuce and tomatoes – own the potato space and work out how to make even more from your specialty).

10. Do as the market manager instructs you.

11. Read the emails sent to you before the market.

12. If you are going to cancel, give some notice.

13. Put serious weights on your tent to avoid customer injury, stock loss and financial and legal costs.

14. Make the most of the opportunity by ensuring you have enough staff to answer questions.

15. Run your own stall whenever possible but if you must send someone on your behalf make sure they are properly incentivised and really know and love your product.

16. Understand how to use social media, keep up with the changes and post pictures of your products with a link to the market. Keep your social media updated with which markets you'll be at and what you're bringing. Share the love by posting pictures of produce from other stallholders too.

17. Markets are hard work. Don't think it will be an easy sell. The instant feedback you get from customers is worth its weight in gold. It is far better than any survey.

18. If a new customer says, "I'll come back next week and buy it" ...it normally means they don't want it and won't come back.

19. Always give discounts to other stallholders. It's the unwritten rule at markets and creates a good vibe, and everyone benefits.

20. Don't ask for discounts on rainy days from the market managers as we don't ask for extra on the sunny ones! It averages out, and that's just nature and farmers' markets.

21. Be patient during set-up and pack-up times and when vehicles are coming in and out of the market. Work with the

market managers and other stallholders to ensure the best results for all.

22. We understand how hard farmers work and that it's not easy, but it makes a massive difference to a stallholder's success if they always turn up with a smile on their face. Smiles and positivity are catching!

Pricing your produce

Pricing your product correctly means that you will get paid for your efforts. Here are some things to factor in when working out whether markets might be a profitable outlet for your produce.

Stall fees

Depending on the market, these might range from $20 to $120, or $400+ for big city markets. Let's say you know that your profit on pickles is $4 per jar. At a $20 market, you would need to sell five jars just to pay your fee (and this doesn't cover your time spent selling the products or your transport costs), while at a $120 market you would need to sell 30 jars just to pay the fee. Fees vary due to the location of the market. Country markets are normally cheaper, while city markets that attract more customers with more money and must pay higher rental fees for the space, charge more.

Cost of samples

For every jar you open, every cheese you slice, every orange you quarter for tasting, there is an associated cost for the product, as well as the cost of the utensils you use to serve the product such as spoons or sticks or cups. Make sure you include these costs in your analysis. Sampling works wonders for sales, but some markets attract customers who want to gorge themselves but have no intention of buying anything. They will literally just grab something from your table and keep walking. If you observe this happening, keep the samples behind the counter and only offer them to customers you think are truly interested.

Bags

In an ideal world, customers would bring their own bags, but we're not there yet so you will need to either pay for bags so customers can get their goods home or pass the cost on to them. You can provide paper or plastic bags, although plastic bags are being phased out at many markets. Bags can cost anything from a few cents each to 50 cents or more. Reusable sacks and bags cost extra. You can also sell reusable bags with your branding on them for customers who have forgotten their own.

Time

If you had to pay someone to do your job, would what you make from the market be enough to do so? Factor in overtime and any other benefits you will need to cover.

Breakages/spillages/theft

Moving products into and out of vehicles, stacking it on tables and hauling it home can lead to breakages and/or damage to labels that might require replacement. Customers might accidentally knock a jar off a shelf or squeeze a tomato too hard, and won't necessarily offer to pay for the damage. Weather events can lead to product damage. Shoplifters might take a fancy to your most expensive produce.

Vehicle wear and tear and fuel costs

The miles add up, and so do the costs of tyres and vehicle servicing. Depending on what's happening around the globe, fuel prices can shoot up overnight. If you are travelling long distances to market, a 20 cents per litre jump in the fuel price can have a big effect on profits.

Stall wear and tear

Bolts pop, materials rip and poles might bend during weather events.

Human wear and tear

Early mornings, heavy lifting, difficult customers, long days and packing and repacking can take its toll on your body and mind. Working weekends might mean you miss out on social events and family sporting commitments. Associated costs can include medical treatment and unhealthy coping behaviours.

Additional market tips

Here are some other tips to ensure that you make the most of your market experience.

Cold storage

- If your product requires refrigeration, it needs to be kept below 4°C. You will need to be able to show officials that you have a process for ensuring the temperature is maintained (such as using a probe and recording the temperature hourly).
- Methods to achieve cold storage could include an esky with ice bricks, 12-volt marine refrigeration systems, and car fridges through to refrigerated trucks and generators. You might also transport the items on ice and then plug a fridge into a power point on arrival at the market.

Trial run

- Do a trial run of your set-up at your farm at least a week before your first market. See how items look when positioned in different ways. Work out how much produce you can fit on the tables and how it needs to be arranged. Set your produce out in multiple ways and then take photos of the various layouts so you can remember what goes where and which layout looked the best.
- Stand three metres back from your stall. How does it look? Are the tablecloths long enough to cover the unsightly mishmash of boxes you've stored under the table? Is it obvious what you're selling?
- Stand to the left and right of your stall. How does it look from those angles? While you have all the items out, experiment with packing them into your car/trailer/truck and take a picture of the best arrangement so it's easy for you

to put the jigsaw puzzle together again when it's time to go to market.

Customers

- Customers bring not just their wallets but also their personalities to markets. You'll be dealing with some of the sweetest souls, as well as the know-it-alls, bullies and cheap-not-cheerfuls. Don't take their personality personally.
- If you have a product that is exceptionally fragile or valuable, put it just out of reach of people's hands, but if you see anyone eyeing it, hand it to them for inspection.
- Indulge in random acts of kindness...offer a customer the opportunity to leave their heavy bags at your stall while they continue to shop for other things, or help them take their bags to their car (if you have someone who can mind the stall).
- If a customer asks to leave an item at your stall and pick it up later (e.g. a heavy item or something in your refrigerator), ask them to leave their car keys so they remember to come back for their goods. It can be such a hassle at the end of the market to discover that you've been left with a customer's forgotten purchases.

Selling tips

- A crowd attracts a crowd. Engage someone in conversation so there is activity at your stall. Move out the front and rearrange items. During quiet times, have one of your staff or family members stand on the customer side of the stall as though they are a prospective purchaser.
- Offer cooking and preparation advice plus recipes.
- Let customers know if your product has certain storage requirements – you want them to have the best experience with it so they come back and purchase again.

- Offer gift packs and suggestions for special occasions such as Mothers' and Fathers' Days.
- Use the seasons to create recipes for in-season produce and to highlight certain product lines, e.g. if you make balms, promote your itchy bite balm heavily in summer and your chapped lips balm in winter.
- Try a product demonstration to attract and inform customers.
- Offer samples to encourage trials.
- If you're new to a market, take note of the early shoppers who are on a mission – these are the buyers you want to reach quickly. Prepare some sample bags and approach them and say something like: "hello, I'm new at this market and I was wondering if you'd try some of our product and let me know what you think of it next time."
- When you have a surplus following a seasonal harvest, create signs that inspire people with all the ways they can use your product. Offer a little recipe booklet for any purchase over a certain amount, for example, if you're selling strawberries, include recipes for strawberry cheesecake, strawberry crumble, strawberry shortbread etc. Or, provide a sheet on how people can preserve the product, e.g. through drying (better still...learn how to do it yourself so you can offer another value added product later in the season when the produce isn't even growing).
- Offer an up-to-date point-of-purchase system that allows customers to use a variety of payment systems.
- Have enough paper bags to package goods for customers.

～

Nuts and bolts

- Keep your cash secure and have plenty of change available. An apron with deep pockets is a safer place to store the big notes from your cash tin than the pocket in your jeans.
- Check the local food laws, but tasters generally need to be behind a sneeze guard, and you will also need a sign reading

"No double dipping". You may also be required to have warm water, soap and paper towels on hand.

- Put together a market kit with scissors, marker pens, tape, a receipt book, business cards, spare screws for your tables and marquee, and a cloth for any spills.
- Always check the weather forecast before you set up so you know what conditions you will encounter. Act on the knowledge gained by preparing for the likely direction of the wind/sun/rain. Shrink your stall by setting your tables further undercover so customers can come in out of the weather and your labels and produce aren't damaged.
- Track your sales over a few markets so you can forecast the supply you will need in future.
- If an item is not selling, try relocating it on the stall, changing the signage or adjusting the price.
- Keep your stall looking fresh, wash the sides down, repair any damage, and repaint and reprint signs as needed.
- It's tempting to buy food at the markets, and that can be half the fun, but if you're on a budget, pack a great lunch and snacks so you're not tempted.
- Ensure that your labels and claims comply with laws.
- If using scales, have them professionally tested.
- When you first start doing markets you might experience a slow start, as it takes time to build traction and loyalty, but don't wait forever – some markets might never pay off. You should have a pretty good idea of whether a market is going to be worthwhile after about four market days.

Finally, be a reliable, pleasant stallholder. Market managers and customers quickly tire of stallholders who miss markets without a genuine reason.

FARMER SPOTLIGHT: Flowers

A former merchant banker and her husband swapped office towers for flowers, discovering passions for horticulture and wildlife that continue to blossom.

Farmer: Lyz Taylor
Farming: Tall Bearded Iris, Louisiana (Water) Iris, Daylilies, succulents and cut flowers
Farm name: KINSpirit Iris and Daylily Farm
Farm size: 15 ha (38 acres) of which just 1 ha (2.5 acres) is cultivated, the remainder of the land being preserved for wildlife
Where: Krambach, mid-north coast, NSW
Average annual rainfall: 1 300 mm
Farming since: 2001

With product names like Behind Closed Doors, Bordello and Covet Me, you'd be forgiven for thinking that Lyz Taylor was selling something other than horticultural produce, but she's not. Lyz is one of NSW's most successful breeders of evergreen perennial Iris and has also won the Royal Horticultural Society of NSW's award for Floral Art.

"As a seven-year-old I was holidaying in the country with my Dad and I asked him why we lived in the city and not here? His reply was, 'there are no jobs.' I remember back then thinking that there must be a way. Now I've finally found it."

Years working in the hustle and bustle of Sydney and London increased her desire for a place in the country. She and her husband John spent a decade taking exploratory road trips north and south, but it was only when John had a serious accident that left him unable to work that they decided to act.

"We spent a weekend looking at eight properties with the aim of finding land that was high and open and within our budget," said Lyz. "Before we'd even climbed half the driveway to KINSpirit, we'd fallen in love."

The couple sold one of their cars to provide the deposit, and Lyz embarked on three years of horticultural training at Taree TAFE college. During the course, she worked for a large flower grower and experienced the effects of drought first-hand, and so set about ensuring water security on the property by fixing gutters and expanding the water-capture capacity.

"I wanted to run a small coastal nursery that was not open to the public, so I focussed on growing my favourite flower, the Iris, which is warm-weather hardy and can be sent bare-rooted by mail order."

Lyz's typical 60+ hour week (35% physical labour and 65% business) involves potting, cutting back flowers and foliage, feeding and fertilising, checking watering systems, putting in seedlings or seeds, selling the plants and paperwork. Lyz sells the Iris and Daylilies via mail order and the internet. She also takes cut flowers and arrangements to farmers' markets, flower shows and gardening events, as well as selling wholesale to selected florists. A new business has also developed out of their regeneration of the land, with the huge variety of bird species attracted to the flowers and trees they have planted now being caught on film and turned into gift cards by John.

"We make money two seasons out of four, and need to rely on off-farm investments to meet the shortfall, but we love this place more than words can say," said Lyz. "Everything about being surrounded by green makes me feel better. Being in this space ignites my whole being. Plus, I get to call my own shots, and you can't put a price on that."

Food for thought from Lyz

What are the basics required for your farming activity?
In the small nursery industry you need to specialise to survive. Find an organisation such as the Iris Society that can teach you all you need to know to get started and help you along the way. It's also great to meet people who have been in the industry for 20–30 years.

What are the threats to the industry?
Climate change. Whether it's in the form of changing temperatures, extreme floods or droughts, you need to select the right things to grow and try to combat the worst effects by creating unique micro-climates on your land to help lessen the impacts.

Opportunities in the industry?
There is growing demand amongst florists and customers for unique foliage (gymea leaves, monsteria leaves, gumnuts etc.) in floral arrangements. I think hardy plants such as Daylilies that can go from flood to

drought and back to flood and still survive have a big future in edible landscaping.

What makes a good farmer?
Someone who's a goer, who makes things happen and doesn't just talk about it. You also need a good backbone because farming isn't for wimps.

Best piece of advice from another farmer?
Feed the plants routinely, not haphazardly.

Worst piece of advice from another farmer?
Someone once said to me, "You need to spend more time online." Many sellers seem to allocate a third of their week to online activities, but I may as well have stayed in the banking industry if I wanted to be cooped up in an office...though I could potentially hire someone to do it for me.

What would you farm if you weren't involved in your current activity?
I'd do more succulents. I've been value adding them in potted arrangements and think they're the most adaptable plants on the planet. They'll cope well with climate change, look great and the honeyeater birds love to feed from them.

Lyz's top 5 tips
1. Love what you do.
2. Have faith in yourself and how you do it.
3. Put a good business plan together.
4. If you wouldn't buy it – don't sell it. Set your quality high.
5. Separate your home and office physically, otherwise you won't be able to switch off. There's nothing worse than the phone ringing at the dinner table and always interrupting your thoughts.

What's the best way to enjoy your product?
Most people just like to look at flowers, but Daylilies are edible, and are full of Vitamins A and C and minerals. Try fresh petals sprinkled on a salad, dry them for soups, or steam the stems and eat them like asparagus. The roots also have a nutty flavour and are great in casseroles.

22

FESTIVALS, FAIRS AND SPECIAL EVENTS

Whatever you're growing or planning to produce, there's a festival or event somewhere in Australia celebrating and selling it.

Think of yourself as hot stuff? Try the Sawtell Chilli Festival in NSW. Like things a bit cooler? Try Dark MOFO in Tasmania. From cider to chocolate, from apples to tomatoes, there's a festival, fair or special event cranking somewhere in the country right now.

To reach the mass market, there are huge exhibitions (with huge price tags to attend) ranging from the Royal Agricultural Shows to big city food fairs, but there are also hundreds of smaller festivals and fairs that are more economical to attend.

There are many reasons to add at least one special event to your diary each year, either as a visitor or stallholder, and some of them are explained below.

Undertake market research

See and taste first-hand how people working with similar produce are innovating and value adding. Markets and festivals are often where trends start. When you're there, focus on stall presentation, packaging, labels, ingredient mixes, taste, pricing and branding. Undercover sleuthing will bring you up to speed with what's on offer, spark fresh ideas of your own and give you a solid understanding of where you currently – and eventually want to – sit in the market.

Discover what impacts

Take note of how produce is displayed, what the sellers are wearing, the signage that stands out, what shopping bags you notice in the crowd and why. Which stalls are pulling the biggest crowds and the most cash? Try to analyse what is working for them and use the information you discover to improve your own business.

Make connections

Meet up with people you might be able to do business with as a supplier or seller. Meet other breeders and growers (perhaps you'll be able to swap knowledge or seeds). If you're exhibiting, get yourself in front of local media and chefs and start building your customer contact list.

Test the market

Test the market while making money. When you decide to invest in a stall at an event, plan well in advance and then use it to test the market for your products, as well as to earn money. By the end of a weekend, if you sell ten times more of one product than another, be grateful that real customers are paying to teach you what works and what doesn't!

Learn from experts

Many festivals offer talks and demonstrations by experts such as chefs and experienced growers. Attend talks and presentations to discover new ways of working with produce, hot trends and what's of interest to customers.

Tax-deductible travel

Where do you want to go? If you are operating a business and this is a commercial trip that satisfies ATO legislation, your trip will be tax deductible. Attending fairs, festivals and special events can offer not only financial rewards but also experiential ones. It's just another element in the "making a living and a life" part of your farming adventure.

Similar to farmers' markets but on a more intense scale, attending fairs, festivals and events is a big undertaking. Long days can lead to long faces, so start small, work out your break even point and see how your fare fares at the fair.

23

SELLING TO CHEFS AND RESTAURANTS

K nives as big as machetes, steam tornadoes, molten aromas, bubbling cauldrons, flame throwers, ice...we could be on the set of Game of Thrones, but we're in a commercial kitchen, where every minute is full of sizzling, scalding action. Does your farm want, or need, some of that action?

While many kitchens spend more time microwaving than starring in the Michelin guide, there are many restaurants and cafés that champion great produce. This is where your farm can come in.

Chefs work in a frenetic environment to strict budgets. The bulk of chefs need the ease of placing orders with a few key suppliers (wholesalers) who they know will deliver at set times so as not to cause even more chaos in the kitchen. Wholesalers might be the main channel via which your produce ends up in commercial kitchens. However, some chefs like to deal directly with farmers who can deliver fresh, seasonal, great-tasting produce. But, be careful what you wish for, as working with some chefs might do your head in.

For example, imagine being called on Thursday morning and hearing "I need 150 perfectly ripe large avocados, each the same size and shape, for a function this Friday. I can only pay $2 per avocado. Can you deliver them to me by 8 am tomorrow?" This is when you need to decide if it is worth the effort of harvesting outside of your normal schedule. Is it worth the time spent on quality control to ensure they are all the same size? Is it worth the one-and-a-half hour return trip to the restaurant? It becomes both a relationship and business decision. Is it worth the effort in the hope that it will lead to bigger and better orders, or would it be better for you to harvest as normal and get a higher rate of return through your other outlets such as the village store, the farmers' market and the wholesaler you have already set a reasonable price with?

Alternatively, working with a chef might give you a core customer to grow for, almost a partnership between kitchen and field. Chefs particularly want to work with farmers growing unique, delicious produce, and they might even feature your farm's name on the menu (e.g. Milly Hill Lamb, Mooral Creek Mushrooms, Burrawong Gaian duck). This stamp of approval can lead to brand building and publicity opportunities for your farm, which in turn enables you to set a higher price for your produce.

~

Tips from chef and restaurateur Guy Grossi, Melbourne, Victoria

What is it that you look for in produce?
It just needs to be fabulous. It needs to have great flavour and it's very important that it's grown in a sustainable manner. There's a food movement away from monocultures, a move toward sourcing product grown with natural fertilisers. I get excited by the idea of permaculture and people growing things that work with other elements of the environment in a natural manner. I also look for passionate people, people who care about animal husbandry and growing food that tastes better and is better all round.

What is the best way for farmers to bring their produce to your attention?
Chefs are constantly researching and scouting for great locally grown produce as it's part of our passion, but you can also contact the restaurant, introduce yourself to the kitchen staff and let the chefs know what

you have that is of interest. There can be a lag time before we respond, but that's just how business is. Great growers and farmers are often in touch with other great farmers and growers, so it's six degrees of separation. I love doing business like this where we have a direct introduction through someone we already know. Networking with other farmers offers lots of potential.

How can a farmer maintain a good relationship with a restaurant over time?

Communication is the key. We love getting the team to do farm visits when we can, and it's all about staying in touch, developing relationships and discussing any issues early in the process so they can be dealt with.

Is there any type of produce you wish was more readily available, or available in a different way?

Victoria is a great source of great produce, and I'd just love to see more people getting involved and experimenting with heritage and traditional methods. I think we need to look backwards to go forwards. As a chef, I want to see food that is virtually ancient. It gives me a buzz.

Any other words of advice for farmers?

Techniques in farming can vary quite a bit. It can be good to travel to other countries where they have thousands of years of experience growing traditional fruit and vegetables. Don't let other people and businesses control what you are selling and how much you are selling it for. Stick to your guns. Great food comes at a cost; that's a commercial reality. Educate people about your farming philosophy, market your produce and please stick at it – we need great people back on the land and we need great farming communities.

If you could only cook one key element of produce for the rest of your life, what would it be?

Tomato. I'd find the best I could, squish them in bottles and use them forever.

∾

Tips from chef Matt Golinski, Gympie region, Queensland

What is it that you look for in produce?
Apart from the obvious things like quality and consistency, I like to know the story behind the produce I'm using. You get a whole new level of understanding and respect for the food you are cooking when you get the opportunity to visit the farms where your produce comes from and develop relationships with the people who grew it. Being able to convey that story back to your customers also gives *them* another level of value.

What is the best way for farmers to bring their produce to your attention?
I think if a farmer wants to see their produce used on menus within their area they should first do some research about who's interested in supporting local producers, then call those restaurants directly (not during lunch or dinner service) and set up a time to drop in samples and discuss the product they are selling. It's a good idea to establish what the chef is already using that might be similar to your product to give them an opportunity to compare price and quality. The things I want to know as a chef are prices, delivery days and times, the quantity they can supply each week, and if it's a seasonal thing, when the season is likely to end so I can anticipate changing to something else. A bit of generosity with samples can go a long way as well, as it shows that farmers are confident and willing to back their product. Good chefs will usually search out producers through farmers' markets, Instagram, Facebook or through word of mouth amongst themselves.

What is the worst way for farmers to try and bring their produce to your attention?
Showing up unannounced in the middle of lunch service and wanting to chat is a bad way to start! First impressions really count, so it's important that my first encounter with the produce and its grower leaves me wanting to pursue it.

How can farmers maintain a good relationship with a restaurant over time?

Good communication is key to maintaining a good relationship between a producer and a chef. Establish a suitable method for communicating that suits both parties right from the start (text, email or phone call) and be sure to keep each other updated on any problems or changes to supply or changes on menus that may affect the quantities needed. It's a good idea to invite the chefs, and even front-of-house staff, out to your farm to show them what you do. Also, eat at the restaurants you supply every now and then so you can see how your produce is being used.

What are some ways you think selling to chefs helps farmers in the short and long term?
Firstly, it's an opportunity to sell directly and get a fair price for your product without going through a distributor. Secondly, it gives you a great platform to build a brand, particularly if the restaurant uses the name of your business on its menu descriptions. Lastly, there's the development of friendships, mutual respect for each other's skills, and direct feedback about your product.

What most annoys you about dealing with farmers?
Nothing annoys me that can't be sorted out through good communication. If a problem arises, it's best to sort it out immediately to avoid either side getting annoyed.

What delights you about dealing with farmers?
Farmers are generally great people who have as much passion for growing food as I have for cooking it. I truly value the relationships I've built with all my suppliers over the years and am grateful for the time they've taken to teach me about what they do. Every time I learn about the way something is grown I think I improve as a chef.

Is there any type of produce you wish was more readily available, or available in a different way?
The variety of produce available to chefs has changed so much over the past 20 years and just continues to grow. I'd like to see more small-scale regional farmers selling their produce directly to restaurants in their area and that becoming a normal part of our industry instead of only being embraced by a few people.

If you could only cook one key element of produce for the rest of your life, what would it be?
Seafood. It's very versatile, fresh and healthy, and I always love cooking with it. But I really hope I never have to choose!

Tips from café operator and caterer Donna Carrier, Bent on Food, Wingham, NSW

What is it that you look for in produce?
It should be local, sustainable and fresh. For bottled products, I look for things that will fill a gap in the marketplace, my shelves or my menu.

What is the best way for farmers to bring their produce to your attention?
Pop in and see the chef or myself and show us what you do.

What is the worst way for farmers to bring their produce to your attention?
Calling us at lunchtime is not ideal, nor is holding the staff up from looking after customers. Be prepared to wait if we're busy.

How can farmers maintain a good relationship with a restaurant over time?
Dine in our restaurant occasionally, let our staff try your products, leave samples and don't start supplying all the cafés and stores close by with the same branded produce.

What are some ways you think selling to chefs helps farmers in the short and long term?
It's a great way to build your farm's profile and over time we can come up with menu ideas together so we can use more of your produce.

What most annoys you about dealing with farmers?

When we're expecting produce that doesn't turn up or is very late.

What delights you about dealing with farmers?
We're able to showcase the region on a plate, and this helps customers appreciate the area's culture and history through food.

What are some negotiation tips for farmers when dealing with restaurants?
Just be yourself and show passion for your produce. When restaurateurs are unable to take your produce immediately, don't write them off. I had an incident where a customer asked me to take his eggs. He sent us a sample, the eggs were great, but I already had an egg supplier and I couldn't take his eggs just yet. He never set foot in my restaurant again, and when an opportunity came up later for another restaurant client I was consulting to I could have promoted his eggs, but we'd lost touch. Basically, never burn your bridges with restaurateurs, as you never know where they will turn up.

Is there any type of produce you wish was more readily available, or available in a different way?
We could always take more fresh produce and a larger range of local cheeses.

If you could only cook one key element of produce for the rest of your life, what would it be?
I love mushrooms, of all types.

Building relationships with chefs and restaurant owners includes understanding how they price meals, how they like to be approached and what seasonal produce they're interested in. As your relationships with chefs develop, you might even grow specialty items for them and plan a planting and supply program with them. This gives both of you some security and leads to less wastage.

Working with chefs can be a highlight of farming. The feedback can be motivational, and it's especially enjoyable to dine on your own produce cooked in remarkable ways. If you want to be a rock-star farmer, start supplying rock-star chefs.

FARMER SPOTLIGHT: Edible flowers and organic vegetables

Leasing land, reaching out to a chef and a passionate desire to grow food were the ingredients this couple blended to create a farming business on Sydney's fringe.

Farmers: Liz and Tim Johnstone
Farming: Edible flowers, microgreens, heirloom and specialty vegetables and herbs
Farm name: Johnstone's Kitchen Gardens
Farm size: 12 acres under cultivation
Where: Hawkesbury, NSW but moving to the Hunter
Average annual rainfall: 750 mm
Farming since: 2010

Liz grew up amongst a riot of colour and flavour in suburban Dee Why thanks to her Greek nanna and a mum who loved flowers and veggie patches. Together with her husband Tim, who has a degree in sustainable agriculture, Liz has seen their produce served in some of the finest restaurants in Sydney including Quay, Icebergs and Bennelong.

"We were renting a house on four acres of land 10 minutes from Windsor," said Liz. "Tim was working full-time and we had no intention of running a farm, but we discussed a few ideas and ended up putting in 1 000 strawberry plants. We had three kids under four at the time, were using backyard irrigation fittings and just kept adding rows and learning as we went."

Liz had no idea how many berries would be produced, and when the harvest kept coming in, she set up a market stall at an arts and crafts market to sell them. About that time, Tim read an article about chef Peter Gilmore working with farmers to get the best produce for his restaurants.

"Tim emailed him, told him our story, that we were a small organic farm close to Sydney interested in growing difficult things and in succession," said Liz. "Peter came out, gave us some seed, we grew it, grew it well, and we've continued from there."

The Johnstones currently lease their land but want to purchase their own plot at some point and employ a team of people. They grow specifically for chefs, as well as for farmers' market customers. Edible garnishes and flowers from peas and broad beans are sold to restaurants, while the unharvested flowers are allowed to mature, with the resulting vegetables sold once a week to consumers at a farmers' market. They also grow unique produce such as salty ice plant, micro and heirloom vegetables, nettles and golden purslane, collaborating with chefs on menu ideas and growing unusual produce to add flair, flavour and texture to dishes.

Liz credits Tim's connections from agricultural college with giving them an entree to a large network of farmers, which in turn gives them access to sources of organic compost and even second-hand boxes.

"It's helpful to cultivate networks that you can collaborate and cooperate with," said Liz, "You can't farm in isolation, and other farmers are invaluable sources of information. Join the Rural Fire Service, and get out and about. You can't just live within your own fences."

Working 60+ hours per week (70% on physical labour and 30% on the business), the work involves picking, weeding, watering, planting, washing, packing, deliveries, staff management and market attendance.

"I love doing the market once a week," said Liz. "Even though I'm up at 2.30 am, it's fun to be face to face with customers and it can make up for the downsides when we have crop failures and weather issues. We've also had to drop markets when the weather hit 47 C and nothing was growing."

"As a family, it would be infinitely more profitable to stop farming and for us both to go into full-time work, but we love doing this, and choose to live our lives without big expenses and farm on a small scale," said Liz. "I get to work from home and have flexibility with the kids, there's something different to do each day and we just love growing. When you get it right, the physical satisfaction of looking at something you've grown – the best you can – is just amazing."

Food for thought from Liz and Tim

What are the basics required for your farming activity?

You need land, whether it's owned, leased or shared. You need tools and access to water, and mostly you need passion and willingness to learn. Farming unusual plants requires observation and intuition, so a lot of agricultural education takes place while you're doing it. Collaborating on social media with other growers also helps.

What are the threats to the industry?

Pricing is the key threat, as the main system for selling food in Australia pits farmer against farmer. This leads to cutting corners and can be very disheartening. For example, we know how long it takes to pick beans, and can't see how the price paid in key markets can even cover the minimum wage.

Opportunities in the industry?

Direct contact with customers.

What makes a good farmer?

Someone who's physically and mentally strong, prepared to make sacrifices and willing to ask questions and incorporate the input of others. You also need to be able to negotiate pricing to ensure that what you are growing is profitable. That can mean growing trial crops first so you can work out the conditions needed, time to harvest, yields and picking times.

Best piece of advice from another farmer?

If you see something that needs doing, do it now. Putting things off doesn't end well on a farm.

Worst piece of advice from another farmer?

We were once told "no spray, no pay", but we have found time and again the 'conventional' way of doing things certainly isn't the only option.

What would you farm if you weren't involved in your current activity?

Poultry.

Liz and Tim's top 5 tips

1. Have a diversified family income base, including off farm income.

2. Be as diverse as you can in terms of your produce, markets, and where you are producing. That is, with produce like ours it helps to have more than one site to spread risk because at times we might have a hail storm, water or pest issue at one site, but not at the other. Different sites also have different weed and pest loadings. Diversity also means that if one product fails or you lose a customer, you still have options.

3. Be as close to the end customer as possible so you can build relation-ships and achieve the best price for your produce.

4. Don't rest on your laurels. There's always a better way to do some-thing. Regular reassessing will get you better results.

5. If leasing land, arrange a commercial lease of at least 2–5 years to protect your venture and make your efforts worthwhile.

What's the best way to enjoy your product?
Either fresh or when prepared by Sydney's top chefs, who take what we've grown and turn it into the most amazing food imaginable, phenomenal dishes that taste even better to us because we know we've contributed to them.

A word of caution. There are many directions you can take with your farming business and it's possible that farmers' markets, online sales or supplying to wholesalers might not be profitable (or enjoyable) for you. It's also possible selling to chefs won't work for you. What can go wrong? You might invest a lot of time building a relationship with a chef who then ups and leaves. Cafes can also take an extraordinarily long time to pay their bills, or go bust and not pay you at all. There is also increasing competition among growers to supply restaurants and if that's not enough, food fashions can change season to season, with last season's hot eat (and which you have tonnes of growing in your paddock), now on the list of not eats.

Like any business, it is important to spread your risk, have multiple sales channels for your produce and always keep your eyes and ears open so you can foresee problems while you capitalise on opportunities.

24

WHOLESALERS, DISTRIBUTORS AND OTHER OUTLETS

Perhaps direct selling to customers isn't for you. Maybe your farm is too far away from population centres to make it worthwhile. Or perhaps you chose to be a farmer because you prefer plants and animals to people. Or perhaps you did the numbers and realised that, in your instance, there was more profit to be had by sticking to your ploughing while outsourcing the selling to others. You might even be thinking of combining direct selling with wholesaling to give your business extra options. That's what this section is all about, opening your eyes to other ways you can get your produce into the hands of customers.

There are many ways to go about this. You can negotiate and sell directly to one or multiple stores, or you can sign an agreement with a distributor who will represent your product in more than one store. If you are growing enough, you might sell to a wholesaler, even sending your raw product to the big central markets. Having your branded produce in a variety of places can be excellent for profile building, but keep the following points in mind when considering selling through outlets other than your own.

You will have no direct communication with the end customer unless you are doing in-store sampling. Because of this, think about having a call to action on your labels, driving people to your website so you can begin to build a connection with them.

Your prices need to be transparent and standardised, that is, sell your goods at the same price to each retailer so there are no misunderstandings or gripes. However, you can set different price levels based on volume sold, which lets you reward retailers who sell more of your product.

Let's say you have an upmarket, artisan product that is hard to sell in your local town. There might be several high-end city delicatessens, health food stores and retailers where your product will be right at home, but you will need to allow for a mark-up of 30–50%, work out the details of who will bear the cost of freight and negotiate payment timing.

Ensure you meet all legal requirements regarding food safety and labelling.

Start small and hone your customer-service skills. Learn how to be a good supplier and stay on top of the accounts so they don't go bad.

There are trolley loads of opportunities out there for good farm produce, including the following.

Food service

In addition to cafés and restaurants, there are clubs, caterers, schools, hospitals and even bulk producers of preserves all needing raw material to work with. Many of these will require you to sell through a wholesaler, distributor or central market. This is because the large food service opportunities often have a head office located far from your fields and military-like logistics and delivery requirements. You might also be required to comply with varying levels of certification including Hazard Analysis and Critical Control Points (HACCP).

Central markets

Did you know that 50–60% of the fresh produce grown in Australia is marketed and distributed through the central market system? More than 15 000 growers supply produce to the country's six market chambers in Sydney, Newcastle, Brisbane, Melbourne, Adelaide and Perth. These growers (and potentially you one day) are represented by more than 400 wholesalers who trade on the floor of these giant operations.

The Australian Chamber of Fruit & Vegetable Industries, trading as

Fresh Markets Australia (FMA), is the national association representing these markets, and they state that the wholesalers at these markets generate more than 12 million transactions a year to retailers, secondary wholesalers, provedores, food processors, exporters and the general public. That's a lot of business, and a lot of hectic activity.

As an example, Brisbane Markets Ltd is set on a 77 ha (190 acre) site at Rocklea in Brisbane. Produce from around Australia is flown, trucked and sent by rail daily to the site, where more than 4 000 people work. Four hundred forklifts move around in a choreographed dance that shifts more than 600 000 tonnes of produce a year.

The Federal Government instituted a new Horticultural Code of Conduct in 2017 that contains requirements for fruit and vegetable growers and traders (wholesalers). A specific requirement is that growers must not trade in horticultural produce with a trader unless the grower has entered into a written Horticultural Produce Agreement (HPA) with the trader.

According to the Australian Competition & Consumer Commission (ACCC), "If you're a farmer growing fruit or vegetables and sell through an agent or to a merchant, there's a law that says you must have a written contract. It's called the Horticulture Code of Conduct (Code). The Code also says that this contract must include certain things, like how price is calculated and when you get paid. Having a written contract protects you. It details what you and the agent or merchant can and can't do. This creates transparency around your relationship with the agent or merchant. The Code also sets out a way for the parties to try to resolve disputes."

The Central Markets are big business, and require thorough investigation if you plan to be selling large amounts of produce.

Wholesalers

A wholesaler is a person or company that purchases large quantities of produce from various farmers. The wholesaler organises the warehousing and logistics required to resell the produce to other businesses such as grocers, butchers, and food-manufacturing companies. In certain cases, a farmer may take on the role of wholesaler for his/her own as well as other farmers' produce.

Distributors

Distributors act on your behalf to get your product into retailers' stores. They normally work with non-competing products. They take care of the initial contact with the store, restocking the store and ensuring that invoices are paid. This allows you to concentrate on production. For this privilege though, you will need to give up even more of your margin.

Supermarkets

Some farmers see getting their product into the big supermarkets as the holy grail for a farming business, while others see being in supermarkets as a 'wholly fail'. Supplying supermarkets requires huge scale and a massive focus on logistics, marketing and quality control.

Playing at the big end of town means you need to be aware that the big chains can change the rules at any time, including swapping your product for that of a competitor, bringing in an identical own-brand product or squeezing your margins so tight that you'll know what it feels like to be crushed in a vice. They can charge you for marketing campaigns and even breakages that happen at their end. You will probably need a Stanley knife to cut through all the legalese and a hole digger to get to the bottom of what will make it a good deal for you.

However, supplying a large supermarket chain will act as rocket fuel for your volume and create huge brand awareness regarding your product.

Organic stores and health food shops

Certified organic produce attracts a premium and is much in demand. Larger-scale certified organic producers often have relationships in place with health food stores, but if you can prove you are farming naturally, they might also be open to working with you.

Butcher shops

These days butcher shops not only sell melt-in-your mouth filos, sweet stir fries and satay sticks, but many also sell cartons of eggs, fresh herbs, pickles, condiments and even fresh flowers. If you think you have a product that might be of interest to your local butcher or one further away, have a chat (preferably while you're buying something and showing your support for their business) and really listen to their response.

You might receive a "no" if they don't have the shelf space, already stock a competing product and are loyal to other suppliers. They might also feel your product doesn't match their brand, have tried something similar before and it didn't work at that time or they have a rental agreement prohibiting them from selling certain products.

Other reasons you might receive a no include they can't be bothered dealing with another supplier, are worried about cross contamination or know the product and the price are not a good fit for their customers.

Finally, they might just have been having a bad day or didn't like the way you approached them because you kept talking while they were trying to serve a customer, or you were too pushy or unenthusiastic. It's hard to get that balance right!

However, you also might receive a "yes"! The only way to find out is by fronting up with a great product, samples, a product sheet promoting and explaining the product, clear pricing (wholesale plus recommended retail price), easy delivery options and a genuine smile.

For livestock farmers, rather than trying to sell their product into a butcher shop, they might even buy the shop! This gives them control along the entire supply chain and into the customer's bag. This strategy of running a retail butcher shop as well as a farm comes with its own challenges, which include but are not limited to competing with the big supermarkets, staffing issues and the limitations imposed by the fact that there are only 24 hours in a day. But, it can also help the viability of a family-run enterprise and offer additional employment opportunities for family members.

∾

Greengrocers

Depending on their size, local greengrocers may get their fruit and vegetables by either heading off to the major markets with a truck themselves or by dealing with buyers at the major city markets. They also work with local growers.

As an example, Ken Little's Quality Fruit & Veg is a grocer based at Port Macquarie, a town of 45 000+ people on the mid-north coast of NSW. Ken's business has about 4 000 retail customers a week as well as 120 wholesale customers ranging from restaurants to corner stores. He receives truck deliveries each day from his buyers at the big wholesale markets in Sydney (390 km south) and Brisbane (570 km north), and also works with local farmers who are trying to sell produce.

Ken grew up on a farm in the New England region of NSW where his family grew peas, potatoes and lamb. One of his fondest memories was as an eight-year-old when his Dad told him that he would bag any of the small potatoes that Ken picked up after harvest that day and sell them at the Brisbane market. Ken scampered about collecting every little potato he could find and eventually filled eight sugar bags. They went on the same truck to the market as his dad's 50 kg bags of larger potatoes, and the next day he heard his dad yelling to his mum that Ken had earned more for his little bags of chat potatoes than his dad had made from the 50 kg bags. That life lesson taught Ken that it's about supplying produce the market puts a premium on, whether it's out-of-season fruit and vegetables, unique produce or produce of such great quality that it's irresistible.

Ken's tips for farmers seeking to build positive business relationships with greengrocers

1. Phone or email first, don't just turn up on the shop floor with a sample, as we might be busy or already have the product or a long-time supplier. Just turning up can lead to an awkward first meeting, which doesn't bode well for the future.
2. The best time to contact us is well before you harvest and often, before you even plant. It's no good turning up on the day of harvest with tonnes of produce to find that we took delivery from another supplier the day before.
3. When you do come in, don't be too pushy or tell us how much better your produce is than that of other growers, let it speak

for itself. Just tell us a bit about how and why you grow, and why you are growing these particular lines.

4. If we're interested, follow up by dropping in with a sample of the product so we can look at its quality and get to know you a bit better. We will also need to know whether it's a one-off supply or will be an ongoing product supply arrangement.

5. Grocers are looking for great quality and a fair price. We are business people too, and know our customers, so just because you are local doesn't mean we can afford to pay $10/kg when we can buy the produce elsewhere at $4/kg. Do your research by looking at the market reports online. They're published each day, and will give you an idea of current prices.

6. Focus on growing saleable produce and understand that you won't get top price for low-quality produce.

7. Get your packaging right. Local growers often have no concept of packaging and try to make do with what they have on farm. This might seem resourceful, but when you're working with grocers, you need to think about how we move, manage and store the product. This means you need to look at new, uniform, sturdy cardboard boxes so they are easy to stack, store and move safely. There's nothing worse than a box's bottom giving way while we're shifting it.

8. Business relationships can grow over time. Start by offering to grow one crop for a grocer, and then if the produce, the communication and the deliveries go well, you will be in position to expand your lines. You might start with spinach and move into cabbages, broccoli, beans and more. If you can plan the crops, you'll be able to deliver fresh every day or so, which means great produce and ongoing supply opportunities. In the case of broccoli, which is often transported in non-environmentally friendly foam boxes for insulation, a fresh supply can be packed into a cardboard bin with a plastic liner.

9. Plan with your grocer so you know how many plants to put in and how often, but we're also open to taking smaller specialty crops as one-offs.

10. A benefit from working with local grocers who also supply restaurants is that we can be your salesperson to chefs. If we know there is some great or unusual produce being harvested soon, we let chefs know so they can plan their menus around it.

11. If I were a farmer, I'd want to deal with a grocer I can trust to be fair. No one should get ripped off. To do that, farmers need to take responsibility for protecting themselves. They can do that by conducting price research at the major markets and by realising that it's no good selling something for $2/kg when it costs $3/kg to produce. Farmers must be able to make money. Not just money to cover weekly outgoings, but the money needed to come back from disasters such as losing $10 000 worth of seedlings in a heatwave or flood.

12. Gaps in the market are often those items that are out of season or affected by extreme weather, so look at season-extending opportunities such as growing tunnels and igloos. You need to understand the limitations though. There's no use growing something there is no market for at the price you want. Do your research and talk to other growers for their insights.

13. Grocers have to resell your produce and make a margin. We also need to operate efficient businesses, so communication is key. Once you are selling to a grocer, be friendly but unobtrusive. Just let the buyer know what you have and when you can deliver so we can plan our orders. If you can't supply us for some reason, we need to know about it so we can get the produce from another supplier that week. By being prompt and proactive, even if the news isn't good on one or all your lines, we can build a relationship of trust over time.

14. Above all, keep all dealings civil. You might get a "no" the first time, but ask the grocer if there's anything else they'd like grown for them. This business is a two-way street.

∽

Tourism-based retail

Tourists are a great market for artisan produce. In addition to sampling local delights, they also often need to take something home as a gift to the neighbour or friend who looked after their pet while they were away.

There are several options to look at when tapping into tourism-based retail including local tourism information centres as well as other tourism operators.

A council-run tourism centre might be open to taking your goods on consignment. Some will require you to be a member of their association, while others might be happy to showcase your produce without a membership.

If the information centre takes your goods on consignment, be sure to keep detailed notes of what you have supplied and when, and what has been paid for because it's easy to lose track if you don't. You will also need to pop in on a fortnightly or monthly basis depending how busy the centre is to restock and manage your display to ensure that it's still looking good.

Another option is to approach other tourism operations in the area. Research local attractions and accommodation operators to get your produce in front of people. If you make goat milk soap, see if you can make small bars for B&B operators. If you make jams from your fruit, can they be served at breakfast time or be included in a welcome hamper? If there is a winery, would they be able to serve your cheese on their platter, your relish on biscuits or offer you a shelf in their tasting area from which you can retail your produce? What about local cruise and tour operators? Do they need any fresh produce to feed the hungry hordes?

Tap into your local tourism network to build relationships that will build your business.

Community stores

Some towns pride themselves on having volunteer-run craft and local produce shops that might charge you a small percentage to have your produce available for sale on consignment. The beauty of these operations is that they are often open seven days a week, so you can direct customers to the store rather than serving them at your farm.

If you want to farm, or grow a farm, there are tens of thousands of potential outlets for your produce. You just need to seed, nurture and harvest the right ones for you.

$\sim$

FARMER SPOTLIGHT: Garlic

A web designer and her husband move from Sydney to Tasmania and win awards for their gourmet produce.

Farmer: Jacquay Christie
Farming: Garlic, olives and blueberries
Farm name: Broadlands Farm and The Tasmanian Black Garlic Company
Farm size: 45 acres
Average annual rainfall: 800 mm
Where: Huon Valley, Tasmania
Farming since: 2011

Jacquay grew up on an organic dairy farm in South Africa, became a web designer and then studied permaculture at Ryde TAFE where she also learnt about holistic farming practices. In 2011, she and her husband moved to Tasmania, and while still working jobs in town, planted 1 500 blueberry bushes, 350 olive trees and a patch of garlic. Seven years later they were still waiting on their first olives and had only achieved limited success with their blueberries, but their value adding to their patch of garlic saw them win two Delicious Produce Awards and the title of Reserve Champion in the Royal Hobart Fine Food Awards.

"Farming is our weekend, weeknight and morning job. We grow organic purple garlic and age it in a fermenting oven for over a month in strict conditions," said Jacquay. "It's long been a part of Asian food tradition, and the end product has tender, sweet black flesh and a texture similar to soft dried fruit. With earthiness and depth, it's a case of sweet meets savoury. It's a premium product, so we package it beautifully to present it well."

The farm requires double the amount of physical labour to business labour, and a typical week involves weeding garlic rows and processing and packing orders. The black garlic sells via wholesalers, retailers, chefs, the Salamanca Market and online, including through aggregator Farmhouse Direct.

In addition to their careers and the farming business, they've also taken on rescue animals including seven horses, two pigs, 12 hens, five guinea pigs, four ducks, four dogs, two cats…and counting. Jacquay finds the death of these animals very hard to take, but it helps that her philosophy is to work hard, be positive and live kindly.

"It's hell to weed garlic by hand, it's expensive to run a farm and you never get time off when you have a city job and a farm job," said Jacqui, "but we've fallen in love with the outdoors and the mountains and never tire of the magnificent views."

Food for thought from Jacquay

What are the basics required for your farming activity?
You need to understand the soils, aspect and rotation practices. Do your research and make sure the crop suits your climate (e.g. we planted four avocado trees in 'hope', but they didn't make it past the first Tassie winter). Any business management skills are helpful. I'm a graphic designer by trade, and this helped with marketing produce, so use whatever non-farming skills you already have.

What are the threats to the industry?
People always want to copy a good idea.

Opportunities in the industry?
If you do it right the first time, with the best you can get, and if you put 100% effort into customer service, you'll be able to jump on any opportunities.

What makes a good farmer?

Someone who truly loves the land, practices biodiversity and is stubborn enough and loves it enough to make it work. A good land steward is someone who looks after the soils, fauna and flora in conjunction with his or her farming interest.

Best piece of advice from another farmer?
"Don't do it!" He was right in some ways because farming is always a game of chasing tails and battling the weather. Despite this, it's very satisfying when you get a perfect crop!

Worst piece of advice from another farmer?
When our fruit trees were getting eaten, we were told that "you've got to shoot the wildlife." This didn't go down well, as I'm vegetarian and bonafide bunny hugger! I learnt that you need to develop a thick skin when living in the country, and that it takes time to adapt.

What would you farm if you weren't involved in your current activity?
I wouldn't farm, I'd just have a bush property and farm trees!

Jacquay's top 5 tips
1. Do LOTS of research – know your crop production methods well, but also experiment because every property is different (e.g. microclimates). What works for others may not work for you.
2. I would say "do it", but have a backup job as well!
3. Look around the world for ideas, as we did with black garlic.
4. Develop a good sense of humour. You will need it for the many times when things go wrong.
5. Be kind to the earth and animals sharing your farm, including the natives, as they were there long before you. Practise ethical, no-cruelty methods if you're a stock farmer.

What's the best way to enjoy your product?
I like to eat black garlic just as it is, or mashed on roast potatoes.

Postscript: the Tasmanian Black Garlic business was sold late in 2017 and is now being run across the bay in Cygnet, Tasmania.

25

EXPORTING AND DAIGOU SHOPPERS

F ancy a frolic in France? A sojourn in Singapore? A boat to Beijing? One of the benefits of exporting is that a portion of your adventures will become tax deductible! More importantly though, according to Austrade, exporting can be a profitable way of expanding your business, spreading your risks and reducing your dependence on the local market. Austrade research shows that, on average, exporting companies are more profitable than their non-exporting counterparts.

Exporting also exposes you to new ideas, management practices, marketing techniques and ways of competing that you wouldn't have experienced by staying at home. All of this considerably improves your ability to compete in the domestic market as well.

Austrade also believes exporting helps you to become more efficient and increase your productivity as exporting companies have better growth prospects, highly skilled, highly productive staff and tend to adapt to technology and best-practice techniques faster.

Interested? You will glean some great information from Austrade, start on their website and then the next step is to get in touch with their network of advisors who can provide assistance in areas including market briefings and cultural tips for specific countries; information on commercial practices and requirements; local industry insights and context; general marketing and promotional advice and opportunities to join trade delegations and fairs. They can also give you referrals to specialist advisors and information regarding potential financial assistance and government grants.

In addition to produce, you can also export knowledge and expertise in areas ranging from animal feed and nutrition to biosecurity services.

The planet has a lot of mouths to feed, so when you decide to export, you're opening your farm to a world of opportunity. Imagine building a global brand from your tiny corner of Australia and providing employment opportunities for other locals.

Daigou shoppers

My first experience with a daigou shopper was when a Chinese lady drove her BMW into our farm's carpark. She warily entered our little farmgate shop and set about tasting the honey.

As she did so, she looked out across the rolling green paddocks (we'd recently had rain so it was picturesque), made lots of "ooh" and "aah" sounds and nodded her head quickly, just like a rooster telling the chooks "the food over here is good." When she said she would like to buy honey, I smiled. No problem, a jar or two, I thought.

"I like to buy two hundred big jars now," she said. "You get?"

"Two hundred?" I repeated, not sure whether I'd understood her correctly.

"Two hundred!"

So, I headed to the shed and got, and got...and got! We even had to pour more on the spot to satisfy her order.

The Chinese word *daigou* means 'buying on behalf of'. Daigou shoppers are trusted recommenders of product. They might have a small network of family and friends in China, or a large network of people relying on them to recommend, purchase and deliver product to them. Daigou shoppers search out genuine product from genuine suppliers. Many large companies now send their product directly to China, but smaller producers of healthy and premium products can work with daigou shoppers as an informal export strategy.

The Chinese market is also well served by several web-based platforms that connect food producers with customers.

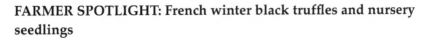

FARMER SPOTLIGHT: French winter black truffles and nursery seedlings

A jet engineer with the RAAF goes from high above the ground to below it with a successful truffle venture.

Farmer: Al Blakers
Farming: Truffles
Farm name: Five Acre Nursery and Manjimup Truffles
Farm size/s: 42 ha (104 acres) and 7 ha (17 acres)
Average annual rainfall: 1000 mm
Where: Manjimup, lower south-west, Western Australia
Farming since: Since he was born, with time away in the RAAF

Al Blakers grew up on his family's farm but joined the RAAF as an apprentice at 15 and then ended up going into civil aviation overseas. Back at home, his father eventually ceased a partnership farming with his own father and brother and started a seedling nursery at the front of the family farm. The market dictated the nursery crops, which included Tassie Blue Gums for wood chips and vegetable seedlings for the vegetable growers in the district. The venture took off, and Al returned to help for 12 months. The rest is history.

A few years later, Al and his father purchased a second farm nearby and although his dad wanted to grow grapes, Al convinced him that truffles were the way to go. He dedicated himself to becoming an expert, and quickly learned that although lots of trees are planted by farmers, and although lots of people think growing truffles is easy, it's only a select few who become viable.

"One of the toughest moments was watching millions of seedlings die right in front of my eyes. It took just a few days for diseases we had no knowledge of how to control to wipe them out," said Al. "Over the past 30 years, we've had to develop many techniques to control these threats and have pioneered many of the control measures commonly used in nurseries today. That experience taught me to never give up, as there is a way around everything."

Al acknowledges that nurseries and truffieres (areas of land where truffles are found) are 24/7 operations and it's difficult to get time off, but his venture has given him the ability to make enough money to indulge in his passion for high-performance cars, speedway racing and Targa Rallies.

Al works 60+ hours per week (25% on physical labour and 75% on the business). Up until a few years ago, he was involved in every aspect of the business ranging from staff management to growing seedlings to all the farming jobs that go with truffle orchards, but he has recently become less physically able due to a hereditary motor neuropathy disorder called Charcot Marie Tooth (CMT) syndrome, which limits his mobility. The farm's produce is sold via wholesalers and retailers, and to chefs. Approximately 90% of the truffle crop is exported, and they also value add by working with a large condiment company to produce a truffle salt.

"I believe I've been successful at farming because I've had a clear vision about what I want to achieve," said Al. "I don't worry about what anybody thinks or says about what I'm doing. I just focus on my objective and let my competitors worry about what I'm up to. Be the leader, not the follower."

"One of the things I love most about how I farm is that I have total control of the growing and selling of my truffles. Nobody can mess with that."

Food for thought from Al

What are the basics required for your farming activity?
To be totally mad helps, as does a thick skin! You also need to have the right land and inoculated seedlings of hazelnuts or oak. Mechanical ability is needed to keep your equipment in good condition and you need green fingers to grow crops. Do your homework, get the right property and don't take any short cuts when establishing the crop.

What are the threats to the industry?
Idiots thinking there's a fast buck in the truffle industry and trying to sell poor-quality truffles.

Opportunities in the industry?
If you understand what the markets require and have a network of associates, you can get the best price for what you are growing, especially as our truffles are harvested at a different time of year to those of European growers.

What makes a good farmer?
Someone who's a jack-of-all-trades and not afraid to take a risk.

Best piece of advice from another farmer?
"If you don't have a go you won't know if it will work." I've found most farmers are reactionary and will usually follow what other growers are doing. Very few will take a chance unless they have somebody to blame if it fails.

Worst piece of advice from another farmer?
It will never work!!!

What would you farm if you weren't involved in your current activity?
Grapes, avocados or beef.

Al's top 5 tips
1. Look for high-value returns.
2. Do your research.
3. Try to be the middleman as well.
4. Don't be afraid to export your product yourself.
5. Listen to the old farmers who have experience, but be sure to find the successful ones.

What's the best way to enjoy your product?
Grate on eggs, drizzle with oil over pasta and risotto, add with cream to dishes and grill with cheese.

PART V

26

AGRITOURISM

I t's possible that *Homo sapiens tourista* is the most challenging species on the planet to farm. Their feed, watering and handling requirements change from minute to minute. They can be unpredictable, uncooperative and downright unmanageable, yet their potential profit per head could make them the most lucrative livestock in your paddock. How do we know this? Because more than 50 000 people have visited our farm over the years. Some stayed for a few hours, some stayed for a few days, some stayed for three months and some have stayed in our memory for years after the hilarious, heart-warming and sometimes horrible times we shared with them.

Agritourism is pick-your-own and native bush tucker mazes. Agritourism is wine tasting and cooking schools. Agritourism is basically any type of activity or offering which brings visitors to a farm.

Agritourism is not for everyone.

The above statement gets its own line to draw attention to its message because if you don't truly love people, engaging in agritourism will distract, depress, infuriate and jetpack your blood pressure into the stratosphere. Even if you do love interacting with people, the demands of dealing with tourists day in, day out will be a huge challenge. So, what are the benefits and risks of agritourism?

Key benefits of agritourism

- Enables you to diversify your income
- Assists in making your farm viable
- Builds your farm's brand
- Extends the opportunity to make money year round, even if your harvest only lasts for a few weeks.
- Gives you direct contact with customers and the opportunity to ask their permission to continue marketing to them (through emails and newsletters etc.)
- Opens the opportunity to receive government funding through tourism grants, development and marketing programs
- Enhances the farm environment
- Allows you to share your love and passion for the land and your produce
- Creates employment for family members and locals
- Inspires new products from your produce
- Creates new networks with other businesses
- Helps you reach and teach people with a special message or philosophy
- Enriches your life through meeting and interacting with a wide range of people
- When you do it well, you'll also get repeat guests and brand advocates, friendship and deep fulfilment.

Key risks of agritourism

- Malicious or accidental damage to your property, plantings and animals (such as gates being left open and animals escaping)
- Too much reliance on tourism can leave you at risk when natural disasters and economic cycles strike
- Biosecurity across your property is harder to maintain

- The expense of making (and keeping) your farm and buildings 'fit for purpose'
- It can take your eyes and hands off your agricultural enterprise
- Injuries and fatalities to visitors (trip hazards, animal encounters, barbed wire, malfunctioning equipment, falling branches, dam drownings, carpark accidents etc.)

There will always be risks with a farm-based venture, but a professional risk consultant can help you to manage and mitigate them. It will be up to you to exercise the attention to detail, the tenacity, and the determination to get past the fear of risk to make your agritourism vision come true.

Are you still keen? If so, there are numerous ways to add agritourism to your venture, or to shape your venture around agritourism, but it helps to focus on the six P's of purpose, product, people, promotion, profits and passion. These are explained in more detail below.

Purpose

This is your 'why?' Why do you want to share your farm with tourists? Why do you want to reach out and be reached? What are you really trying to achieve?

You need to be clear on your purpose so you will be able to sustain the attitude and energy needed to be successful.

Product

What is the product you offer, or plan to offer? Think of it in terms of both a physical product and a tourism product.

Physical product includes buildings and facilities, land use and size, livestock, general appearance of the farm, topography, vegetation, pasture, soil types, water and water features, location and access, produce.

Tourism product includes experiences, services, accommodation as well as how these experiences and services are delivered. It includes the factors that make your farm unique and the behind-the-scenes planning including agriculture, risk management, maintenance and workplace health and safety that enable you to deliver the experience

People

Who are your people and what role might they play in your agritourism venture? Will it just be you and your family? Or might you have employees or entrepreneurs working with you on your land? What qualifications and qualities will they need?

'People' also includes your neighbours. When you are on good terms with your neighbours, everything is smooth sailing, but when you're not, it can be a busload of trouble. Be sure to keep communication open with your neighbours so they are positively impacted by your foray into tourism. Perhaps you can offer employment, friendly meals and get togethers and other perks so they too can enjoy the benefits of your tourism enterprise.

Finally, 'people' are your customers. Are they domestic or foreign tourists? Singles, couples, or families? Travelling independently, or with a tour group, or with their pet? What are their interests? Why will they want to come, how will they want to be treated and how much will they be prepared to pay?

Promotion

What is your unique selling proposition (USP) and how do you intend to reach your customers?

How will you use social media (e.g. Facebook, Instagram, Pinterest), rating sites (e.g. Trip Advisor, Google Reviews), share-economy sites (e.g. Stayz, Airbnb) and traditional media to spread the word?

What local, state and national groups can you join to tap into extra resources, support and learning?

Profits

How much money will you need to take in to ensure a profitable enterprise? How will you ensure a good return on capital? How will you stay cash-flow positive and on top of your debt? How will you grow your wealth?

There are many extra costs (both in dollar and time terms) that need to be factored in when you morph from agriculture to agritourism. Some areas to consider are the following.

- Initial compliance with local, state and federal laws. This could involve the cost of making facilities disability friendly, paving carparks and applying for development approval. You might also be required to pay for licenses and inspections.
- Marketing costs. Logo and brand development, website creation and maintenance, advertising costs and launch promotions.
- Interpretation. Preparation of signage and delivery of interpretation in a way that tourists will enjoy (e.g. printed handouts, sign boards with QR-coded links to web pages, downloadable audio, guided tours).
- Directional signage. You will need to work with the roads department to ensure guidance and safe entry to the farm, including appropriate early signage.
- Day-to-day costs. Cleaning, electricity, water, linen, breakages and staff.
- Insurance. A little word which can cost so much. Be very clear with your insurance agency regarding the activities you will be offering to tourists, otherwise they will find it easy to avoid paying out if something goes wrong. Insurance companies are particularly wary of farm-based activities involving archery, horse riding and water.
- The costs of making your farm fit for purpose including the implementation of policies and procedures. You will need an evacuation plan including an identified assembly point, smoke detectors and exit signs and emergency lights in buildings. Hot water temperature needs to be regulated so there can be no scalding. Waterways might need to be fenced and branches at risk of dropping removed. You will need disabled access, plans for moving people about the farm and procedures regarding everything from not leaving the key in

the quad bike to how often the paths need to be cleaned so they don't cause slips.

Passion

It is your passion for your farm, produce, animals and fellow humans that will really resonate with visitors. What are some ways in which you can demonstrate this passion to your customers? Is it the way you physically present your farm, animals and products? Is it the way tours are delivered? Is it the way you treat the earth, staff, livestock, plants and people?

Passion can make up for shortcomings in other areas, but without it you won't be able to carry off the huge undertaking that is tourism – even if everything else is in great shape. Passion helps you remain resilient and positive, and keeps you moving forward when others would give up.

Sometimes the burden of legislation and compliance can cause people to give up before they even get started, but if you approach the challenges methodically – and economically – they can be overcome.

In your planning, work out the above set-up expenses and then add in ongoing running expenses to determine whether what you will need to charge to make a profit is in line with what customers would be prepared to pay. If you are targeting low-spending tourists, you will need a whole lot more through your gates to make it worthwhile, whereas if you are chasing high-end customers, you will get by with far fewer visitors. So, what are tourists are willing to pay for? The following chapters provide an introduction to the key areas, namely, accommodation, experiences and education, agritainment, and functions, conferences and corporates.

～

FARMER SPOTLIGHT: Agritourism

With a background managing resorts and a degree in animal husbandry, an agritourism venture was the perfect choice for a couple wanting to raise their kids in the country.

Farmer: Andrew Campbell and Anna Featherstone
Farming: Tourists, ideas, animals, bugs and plants
Farm name: Honeycomb Valley Farm
Farm size: 36 ha (89 acres)
Average annual rainfall: 1 200 mm
Where: Nabiac, NSW
Farming since: 2006

Getting out of Dodge and becoming more self-reliant were the key aims for Andrew and Anna (yes, that's us!) when they moved out of Sydney onto a rundown farm just outside the tiny village of Nabiac. They picked the area because of its proximity to their family in Sydney, the climate, handiness to beaches and because it was in a price range they could afford. The property also had a great aspect and an underdeveloped farmstay business that Andrew knew he could improve. The potential for the farmstay was key, as the couple had no off-farm income or chance of employment in the area, and the soil didn't lend itself to easy farming, though they've progressively improved it over the years.

"We approached the farm as an entrepreneurial passion project," said Andrew. "We pursued all our personal interests and tried to work out a way to make money from them. Half the fun was having a life plan but not an exit plan, but commercially we wouldn't recommend that for others."

When they first moved to Nabiac, the couple joined Landcare and Seedsavers, read every book under the sun on subjects ranging from alpaca training to small-scale grain-growing, and began to build the business, which went on to win state and national tourism awards. Anna was also mentored in keeping honey bees and native bees and offered courses at the farm in beekeeping, balm and soap making.

"We hardly spent a cent on marketing, but put a huge effort into guest experiences and customer relations," said Andrew. "This meant word of mouth, and repeat business really took off, including some guests who returned more than 20 times and who became great friends."

Working 60+ hours per week (50% on physical labour and 50% on admin/management), the couple – with the help of backpackers for many years – hosted farmstay guests, bus tour groups and day visitors. On a small scale they produced goat milk, honey, beeswax, turmeric, herbs and rosellas (wild hibiscus). These were value added into products including soaps, candles, balms and foods. They outsourced condiment-making to a chef, herbal tea-blending to a professional and worked with other local artisans to value add their produce. They sold online, at farmers' markets, through a farmgate shop, through tourism outlets and through the local volunteer-run community shop.

"We had to work incredibly hard, but were able to make a living over more than a decade while experimenting, meeting people, improving biodiversity and growing personally and professionally," said Andrew. "It was chaotic at times, crazy at others but incredibly enriching and fulfilling."

"Extreme physical work did come easily the first nine years, but part of life as you get older is realising that it doesn't necessarily make sense to keep doing so. Recently we've backed off and I'm loving working in small business and agritourism consulting."

Food for thought from Andrew

What are the basics required for your farming activity?
An entrepreneurial spirit, a web strategy to attract tourists and an offering and level of service that makes people want to come back.

What are the threats to the industry?
Downturns in the economy can affect tourism spending. The cost of insurance and compliance continue to rise, while the climate can play havoc with farming and outdoor activities.

Opportunities in the industry?
Agritourism is just getting bigger and bigger. It's an industry that welcomes individuality and innovation, but you need to approach it professionally if you want to stand out from the pack.

What makes a good farmer?
Someone with problem solving skills, attention to detail and a can-do attitude. If you're working in agritourism, you also need a radar for risk along with good doses of kindness, patience and energy.

Best piece of advice from another farmer?
"Go for a solar pump rather than a petrol one." The solar pump we put on the dam shifts a huge amount of water around the farm to troughs and irrigation and it just keeps on keeping on.

Worst piece of advice from another farmer?
"Farm meat, it's the only way to make money." We didn't want to be in the business of selling animals for slaughter, so used ultra-cute Miniature Galloways as grass cutters and manure providers. They were hardy, had a great temperament and so beautiful that people wanted them as pets. They were also such luscious eye candy we used their pictures to help market the farm.

What would you farm if you weren't involved in your current activity?
I think we'll always farm ideas.

Andrew's top 5 tips
1. Make the most of chore and travel time by listening to inspirational and informative podcasts.
2. Set clear boundaries so guests and staff don't intrude on your down time.
3. Take pride in being kind to your animals, enriching the soil and running a great business. This will show in online reviews and return business for your products and experiences.
4. Build your brand and find customers prepared to pay what your product is worth.
5. Never forget why you started on this adventure in the first place.

What's the best way to enjoy your product?
You're reading one of our products now, hopefully in a comfortable position with a pot of herbal tea or something chilled!

27

ACCOMMODATION

Providing overnight accommodation options on your farm is one way to launch yourself into agritourism and increase your revenue more quickly than ever before. More quickly, because thanks to the rise of sites such as Airbnb and Stayz, not to mention social media, you can easily promote your farmstay to domestic and international travellers. Once you have your first few glowing testimonials, it's easier to get the next 10, 20, and 30 as word of mouth builds and the trust in your operation grows.

Your farm's location will sometimes determine the type of customer you can attract. On a station in the outback? You'll likely be targeting more caravanners than honeymooners. In wine country? You might see more hens and wedding groups than school groups. A few hours out of the city? You might set up for families or the romantic weekend away market. Close to the city? Perhaps you'll be targeting the Asian tour-group market.

You can purpose build for the market you are trying to attract (budget, luxury or something in between), or make the most of what is already on your farm. Our family bought a property that already had a farmhouse and two simple cottages on it, but if you are starting from scratch there are all sorts of ways to get started. For instance, are you using all the rooms in your farmhouse? Is there an old dairy on the property that you might be able to do up or a patch of grass down near the forest perfect for a tent?

Here are some examples of the different types of accommodation you could look at providing.

- Camping, where you either rent out swags and tents or people bring their own with them (take extra care when designating campsites as they should not be at risk from falling branches, flash floods and other hazards).
- RV and caravan sites for people touring the countryside.
- Glamping, such as Tepees, Mongolian Gers, Yurts and well set up tents (even see-through bubble tents!) on platforms.
- Bunkhouses and large undercover areas for school and sporting groups.
- Sustainable accommodation such as straw bale, mud brick or Earthship (made from re-used tyres).
- Authentically and rustically renovated farm buildings such as barns, sheds and dairies.
- Converted shipping containers.
- Permanent caravans, buses or train carriages.
- Pre-fabricated cottages.
- Purpose-built cottages or lodges.
- Part of your farmhouse run as a B&B.
- The entire farmhouse or a purpose-built farmhouse.
- Accommodation within a natural setting on the property such as a cave, a tree house or a houseboat on a dam.
- Unique buildings where your imagination shows through such as a rotating farmhouse, a glass-bottle building or an underground house.
- Kennels, catteries and/or stables to run a pet resort.

If you build it, they will *probably* come, but will they come again, or tell others to? They will if you are an outstanding host.

It will be your top-notch customer service skills, empathy and enthusiasm that will make the difference. Your strong mix of courtesy, patience, professionalism, thoughtfulness and humour. Your ability to make people feel comfortable, listened to and welcome. Your uncanny knack of sensing what's needed before being asked and spending more time using tact than your tractor. People respond to authenticity in others, but in the era of online reviews, if the authentic you is a grumpy, selfish whinger, it's probably best you don't set yourself up to fail.

Running accommodation on a farm means you are running a mini-motel of sorts, and like any hotel-type operation, you will be operating in the following areas.

Reservations. Customers these days are impatient, and if they call in the morning and you don't get back to them until the end of your farming day, they'll have already booked somewhere else. A simple online reservation system helps to avoid this, and there are numerous versions on the market. At the time of reservation, you could forward all the information the guests will need for their stay including a detailed list of the items (e.g. frying pan, wok, full-sized fridge, high chair etc.) that are provided in their accommodation. Having this information on a single sheet of paper will save you answering the same questions again and again. It also provides you with a stocktake list so you can check if anything is missing or broken when guests leave.

Housekeeping, cleaning, hygiene, laundry. Is linen included in the price or do they need to bring their own? What cleaning standards will you and your guests be happy with?

Supply management. Consumables such as toilet paper, detergent, matches.

Front office and guest relations. A friendly and informative welcome. Suggestions for where to visit in the area, either verbally or in a printed compendium.

Finance. Managing costs and budgets, upfront deposits and final payments.

There are as many ways to operate a farmstay as there are strains of wheat. A key question to ask yourself is *what level of hosting do I want to provide?*

None? You can leave a key in a coded key locker at the front gate, a welcome manual in their accommodation and potentially not set eyes on the guests all weekend. Be prepared for phone calls though, as often what you don't think of as an issue, such as lighting the fireplace, turning on the TV, bird calls at night, might be totally incomprehensible or unnerving for your guests.

A simple B&B? This is where you welcome guests, leave them with breakfast supplies or cook their breakfast, and let them do their own thing for the rest of their stay.

A hosted farmstay? In addition to your meet and greet, you have daily scheduled activities the guests can participate in such as animal feeding rounds, property tours and campfires.

If you are in a region that already attracts large numbers of tourists, such as an area renowned for vineyards, the addition of accommodation on your property will help you to tap into the tourist activity all around you. You might not even need to spend much time at all with your guests if they are staying at your property just to be near other attractions. However, if your farm IS the attraction, this is where the accommodation, your hosting skills and the activities you offer need to be real crowd-pleasers.

In addition to humans, who else might you be able to accommodate? You could establish a boarding kennel and cattery, a horse hotel, pet resort or create a farm animal sanctuary where you are paid to care for other people's favourite animals when they can no longer do it. Another way of creating income might be to care for a local farmer's pets and livestock, at their place, while they go on holiday.

28

EXPERIENCES AND EDUCATION

F arm experiences and education can be offered whether you are providing overnight accommodation or not. You can offer experiences daily, on weekends, seasonally, or as a series of events. You can target a wide range of audiences including the following.

- Groups and clubs including special-interest car, gardening, religious, Probus, sporting, parenting and seniors clubs, as well as mainstream school groups, respite care and tertiary agriculture students.
- Tour groups travelling on organised tours.
- Free and independent (FIT) travellers are those holidaymakers who are travelling independently of organised tours. They can be locals from your region or from as far away as Rajasthan.
- Children's parties to wedding groups.
- Specialised experience as well as information seekers such as foodies, people new to sustainability, new farmers, birdwatchers and photographers.
- Conference, seminar and motivational groups.

Animal experiences

Among the most common experiences on offer at farms are activities to do with animals. These might take any of the following forms.

- Petting zoo. A focus on safety, hygiene and animal welfare issues is key to success. The costs associated with a venture of this type include those related to veterinary care, feed, staff, fencing, shelter and insurance.
- Milking. Providing the opportunity for visitors to watch or get involved in the milking of animals.
- Rides, be they on horses, camels or in a cart pulled by oxen.
- Mustering or cattle and sheep work.
- Native animals. The opportunity to experience native animals in their natural environment, whether it be koalas, wallabies or glow worms!
- Birdwatching or night spotting of animals.
- Niche experiences such as pony painting parties, bunny rabbit walking or pig races.
- Training people how to work with animals such as horse whispering classes, alpaca handling techniques, grooming cattle and chickens for shows, teaching pigs tricks and sharing how to harness billy goat carts.
- Animal therapy and social farming sessions. This is where animals and farm experiences are used to help heal, build trust and confidence, break down psychological barriers or assist with therapeutic exercises. It could also be used to rehabilitate people involved with crime or addiction.

Food and beverage experiences

Food tourism is booming, and there are many ways to offer an experience to visitors.

Produce tastings and sales

You can offer this on-farm in your shop, at local markets, or at the tourist information centre.

On-site cooking school

Whatever you are growing, a cooking class can be built around it. Bread making, cheesemaking, fermenting, home butchery and sausage making, seafood preparation, the use of herbs and salad dressing techniques or dessert making –there is a mouth-watering array of masterclasses that budding chefs will pay to attend.

If you're better at digging up asparagus than slicing it, perhaps you can engage a local chef to run the courses for you, or bring in experts from around the country to create a calendar of monthly events.

Plan to make as much money from the event as possible. This might mean becoming an on-seller of specialised equipment (such as cheesemaking kits), required take home ingredients (such as spice packs), and even aprons and recipe books.

The Agrarian Kitchen in Lachlan, Tasmania is a great example of this. The 5 acre farm has been offering cooking classes since 2008. The brainchild of Severine Demanet and her husband Rodney Dunn, the couple expanded their enterprise in 2017, establishing an eatery and store 5 minutes down the road in the old mental asylum building in New Norfolk.

Rodney has great heritage for this type of project. He grew up on a farm in rural NSW, apprenticed as a chef and then moved into food media, including working as an editor at Australian Gourmet Traveller Magazine. He also produces cookbooks.

The couple's love of great produce has seen the five acre farm transformed into an organic food lover's dream with extensive heirloom vegetable gardens and 180 metres of wire trellising supporting raspberries, boysenberries, loganberries, marionberries, silvanberries, tayberries, youngberries, gooseberries, blueberries and red, white and black currants. An orchard provides an array of fruits ranging from heritage apples to quinces, while long term projects include the establishment of hazelnuts, chestnuts and Franquette walnuts.

The cooking classes on offer often feature celebrity chefs and cover a mouth-watering array of specialties ranging from charcuterie, smoking and preserving, to cooking over fire, pasta making and pastry. Private classes are also offered for corporate team building exercises and incentives as well as for groups of friends.

Appreciation nights

Appreciation nights are a way to create deeper bonds with regular customers, introduce new customers to your products and spur on sales. Share an indepth knowledge of how to taste and appreciate your produce, whether it be wine, beer, honey, cheese, tea, or even different varieties of apples. Discuss tasting wheels, aromas (green, grassy and fresh, or berry or nut-like), viscosity, colour and clarity. Demonstrate how sublime it can be to experience the best food-and-drink pairings. If you do a great job and attract the right attendees with the right amount of income to splurge, they'll be leaving laden with both knowledge and armfuls of your produce.

Pick your own operation

People love getting their hands on fresh apples, berries, cherries, citrus, nuts and flowers, and even watermelons and pumpkins. In areas thriving with orchards, pick-your-own fruit and nut operations attract thousands of visitors each weekend in season. This influx of people needs to be managed, so you might need to install special fencing to keep people away from areas you don't want touched. You may need to provide 'training' sessions so people know how to pick the fruit without damaging your plants, and you may need extra staff to help out. You will also need to determine the best way to be paid for what you have grown, for example you might charge by the bucket or by the kilogram. Whatever it is you are selling, think about how you can up-sell customers so they also leave with something else. If you're selling cherries, it might be a fresh slice of cherry pie, or a packet of dried cherries, or a recipe book you've prepared so they know what to do with all the goodies.

On-site restaurant, café or food truck

There is something about the human stomach that needs filling, which is why some farmers choose to fill a gap in their area for farm-based food outlets. Perhaps you want to run a simple café serving homemade pies and milkshakes, opening for just a few hours each weekend. Or maybe you are planning to build and outsource restaurant operations to an enterprising chef. Maybe you already have a food truck that travels to events serving your famous farm grown chillies in a chilli con carne that people rave about. If so, perhaps you might also host a monthly Fiery Chilli and Fire Pit night at your farm. There are many ways to approach on-farm food operations, and a lot will depend on the surrounding population and potential customer base.

On-farm dining events

For an on-farm dining event, you don't necessarily have to have a fully functioning commercial kitchen and restaurant on site.; you can bring in the equipment and marquees or host a long lunch or harvest feast in your paddock. Work with local chefs and other farmers to provide produce to the event and promote it. Offer diners the chance to see a part of the farm before they sit down to their feast. A long lunch, or 'Meals in the Fields', as one of our local farms, Near River Produce calls their events, attracts foodies from far and wide. It creates excitement, publicity opportunities, brand building and, if done smartly, a nice pop in income.

Specialty events

There's nothing like a grape stomp, passata bottling day or watermelon-carving demonstration to get people out to a farm. What might work at your place and for your produce?

Arts, crafts and skills

Farms provide a backdrop for craft and artistic pursuits as well as skills training, not just because of the potential landscape, but also because of the natural materials that are present on site. Areas of opportunity could include the following.

Natural-dyeing workshops

Using natural dyes from plants such as berries, bamboo, barberry and so many more, you could run interesting sessions teaching people how to colour materials and fibres naturally.

Natural-building workshops

Invite others to work alongside you as you make structures from cob, mud bricks and straw bales.

Photography lessons
Rusted farm machinery, cobwebs shimmering at dawn, animals in motion – these are great subjects for a photography class at your farm.

Art classes
Picasso in the paddock, watercolours by the water tank, charcoal and crayon amongst the cattle. If your idea of art is stick figures, you could invite local artists to host the classes. An interesting idea for beekeepers is to have someone teach encaustic painting, an art form using heated beeswax with pigments applied to timber and other surfaces.

Indigenous knowledge
Work with local indigenous groups to offer bush tucker and cultural experiences.

Dreamcatchers
These are made from twigs, feathers and other natural materials foraged from the farm.

Sculpture
Use items lying around the farm such as corrugated-iron roofing, barbed wire and timber, or natural resources such as clay to help people create interesting sculptures.

Woodwork
Pass on your knowledge of whittling, or making musical instruments, cutting boards or toys.

Restoration and maintenance of farm equipment
Help people to learn the skills they need to maintain common farm equipment from chainsaws to tractors.

Blacksmithing

Get that fire burning and teach the old art of blacksmithing, encouraging your students to make knives, jewellery and decorative horseshoes.

Spinning, weaving, quilting and felting
Using fibres from the farm, spin up some new business.

Tanning and leatherwork
What a hide to suggest that you and others get their hands dirty tanning their own leather, but if you think about it, what an aspirational course this could be for a city person to be able to say they created their own floor covering or handbag.

Bushcraft
Tap into people's fears and aspirations by teaching them survival skills ranging from foraging to creating fire, from capturing water and making it safe to drink, to hunting and building shelters.

Farm skills
Teach others who are new to farming how to build, repair and maintain fences, use a sawmill or grow specialty plants.

Drumming
Hold a drumming workshop or teach people how to make their own drums from animal skins.

Writing
Host a writers' retreat.

Flower arranging
What a lovely way to spend a morning, picking flowers and crafting them into beautiful arrangements and bouquets.

Scarecrows
Hold a scarecrow-making workshop as a fun school-holiday activity, and then use the completed scarecrows as a seasonal art installation.

Gardening workshops

What are you good at and what do people want to learn? Hold a seed-saving Saturday or a session on organic gardening. Teach people about local weeds or good predatory bugs.

There is a thirst for knowledge in the community, and you could help sate it.

~

FARMER SPOTLIGHT: Stockhorses

Breeding and training Australian stockhorses, running horsemanship clinics and competing in and judging the unique Australian sport of campdrafting keep Vicki Evans either up in the saddle or on her toes.

Farmer: Vicki Evans
Farming: Australian Stockhorses
Farm name: Talawahl Australian Stock Horse Stud
Farm size: 80 ha (200 acres)
Average annual rainfall: 1 200 mm
Where: Nabiac, mid-north coast, NSW
Farming since: she was born

Many little girls dream of owning a horse, but Vicki Evans has taken it to the next level with at least 50 Australian stockhorses in her paddocks. Vicki's love of all things equine was established at her family's dairy where she rode horses for chores and discovered a natural affinity with them. As an adult, however, she soon learned that horse farming is not just about breeding lots of stock. It's about producing horses to meet the market need, whether that be no nonsense horses or horses for performance work.

"The Australian stockhorse is an amalgamation of a variety of breeds including thoroughbred, Arab and pony," said Vicki. "They are generally between 14 and 16 hands, they can be light or heavier built and they're known as the 'breed for every need'."

Those needs can range from riding for pleasure to dressage, polo, polocrosse, show jumping, eventing, pony club, working, hack and campdrafting.

So, what's campdrafting? It's a uniquely Australian sport, the aim of which is for the horse and rider to work together to move cattle around a set pattern on a course in a timely and visually pleasing way. It's a big family sport in the country, with lots of tree changers getting on board. Vicki loves competing in and judging the events, as well as training people for them. She keeps 30 head of beef cattle on the property to complement the training and education of the horses, as well as for pasture management.

"Two plus two doesn't always equal four when you're breeding," said Vicki. "Just because you put a great stallion over a great mare doesn't mean you'll always get great progeny. You select along the way for consistency, and it can be tough in the early days when you need to cull animals who might present a danger to riders. There can also be heartbreak if you lose a good horse to an accident."

Working 60+ hours per week (90% on physical labour and 10% on the business), Vicki's work involves all aspects of horse husbandry and organising her stallions to service customers' mares, as well as training horses in preparation for sale or competition. She is increasingly running clinics and coaching sessions. She sells her horses through word of mouth, and appearances at sales and competitions. She also advertises her clinics and coaching on Facebook.

"A lot of the work of a horse breeder is matching horse and human. It's important to get that right so you can sleep at night," said Vicki. "It makes me so proud when someone recommends me by saying 'any horse you get from her will be alright', but it still means I need to steer people away from horses that aren't right for them."

"It can be a hard, physical slog and my jeans practically walk around on their own at the end of the day," she said, "but I love the result of seeing what I've produced go on and do good things."

Food for thought from Vicki

What are the basics required for your farming activity?
You can't wake up one day, look through a glossy horse magazine, buy a stallion and just jump in. It takes a long apprenticeship to run a

successful stud, so get as much hands-on experience in the industry as you can. There are also clinics and horse husbandry training to help you on your journey, but if you're not a horse person, they won't make you one. When you are establishing yourself, remember that this business is all about reputation, so you need to put up horses you are proud to sell.

What are the threats to the industry?

In some years there is an oversupply of horses. Gluts occur when people get into breeding thinking it's easy money. It's easy to get a hold of genetics and get foals on the ground, but there's so much work involved and it takes years to train horses, build a stud's reputation and sell consistently.

Opportunities in the industry?

Many people buying horses don't have the skills, time or experience to train and work their own, so they're looking for quiet, no-fuss, easygoing horse who is good after a spell and won't bolt or buck. There's a market for calm, trainable, forgiving horses that are willing to perform and work with the rider. That's what leisure riders need. There can be no nasty attributes like kicking, biting, bucking, or rearing.

What makes a good farmer?

A horse person needs to be calm, tolerant and patient, with a genuine love of horses and an appreciation that each horse is different. A good farmer is always learning, and you need to keep your sense of humour handy. Gary Larsen cartoons are good for that!

Best piece of advice from another farmer?

"If you give yourself all day to achieve something with a horse, it won't take long. But if you give yourself a time limit, it will take all day." My old mate was telling me to allow time and be patient. Another great piece of advice was to "swear politely and calmly."

Worst piece of advice from another farmer?

Kick it harder.

What would you farm if you weren't involved in your current activity?
Burmese kittens.

Vicki's top 5 tips
1. Love what you're doing and know what you're doing.
2. Have a partner with an off-farm income.
3. Skilled horses don't come cheap and cheap horses don't come skilled.
4. Breeding horses isn't about quantity, it's about quality. You don't want to overbreed or produce poor quality horses that need to be culled.
5. Don't get a foal and have it grow up with kids. They're not puppies and someone will get hurt. Things can happen with horses when you least expect it. You need a lot of common sense, but unfortunately that's not as common as it used to be.

What's the best way to enjoy your product?
Horse is lovely satayed. Just joking! We have a forest out the back, and I don't care how late in the day or how tired I am, I feel my shoulders relax when I head through that bush on horseback.

Physical fitness, water and adventure

Farms provide an opportunity for all sorts of active adventures. There will be safety risks and insurance premiums, but adventure activities can bring a whole new group of customers to your property and could lead to repeat custom from locals too. Here are just some ideas.

Farm boot camp
Fitness instructors can use your farm to motivate and muscle up their clients. From moving hay bales to digging over garden beds, herding animals on foot to shifting firewood piles, this type of functional fitness leaves everyone feeling good.

Segway tours
Charge people to explore your farm on a Segway.

4WD tours
Excite guests with a ride over rugged country to a beautiful vista.

Fishing
Let guests try to catch the silver perch you've filled the dam with, or try their hand at catching bass, river fish and yabbies.

Kayaking, rafting, canoeing, pedal boats and sailing
Make the most of your waterways.

Inflatable waterpark
Add fun, temporary structures to your dam during the summer months.

Hovercraft
It's a level up from a hay ride, but how about charging people to hover across your paddocks and float over your dams?

Adventures up in the trees
Professional treetop obstacle courses continue to attract fans.

Abseiling and rock climbing
If you have the topography for this, it's another activity to rise to.

Mountain bike riding
Bike riders will love the opportunity to try and stay on track.

Bushwalking
Provide maps and paths for nature lovers to take a hike.

Archery
This can be tricky with insurers, but is an attractive activity for some visitors.

Simple activities
Whip-cracking and boomerang throwing are fun to teach and learn.

Drones

Restrictions mean that city people don't often get the chance to fly drones, but on a farm with wide open spaces there is ample opportunity for them to learn.

Bike riding tours
Join with other local farms to offer a fully provisioned, self-guided three-day bike exploration of your region with overnight stays at different farms.

Quad bike adventures
High risk, but potentially high return, you might be able to rev up interest in your farm by offering visitors quad rides.

Renewal and wellbeing

Do all those other activities sound too hard, too energetic and too much? Slow the pace down with a focus on mind, body and spirit.

Yoga and stretching classes
These can be held in a tranquil location on your farm, or if you want to add some fun, bring in the baby goats to join in a goat yoga session.

Seasonal celebrations
Bring people together to appreciate the full moon, solstices and equinoxes.

Healthy eating
Help people discover raw foods with picking and cooking straight from the garden.

Meditation
Introduce people to the art of meditation and mindfulness.

Relaxation
Provide people with the chance to do nothing but take in the view.

29

AGRITAINMENT

A gritainment is agritourism on steroids. It's getting people to your farm even if the link to agriculture is less than tenuous. It's about making the most of your facilities and getting the most out of your customers' wallets. There will be numerous permissions and regulations you'll need to deal with before you can offer this type of farm-based tourism, but here are some ideas to get you thinking.

Concerts, theatre and movies
Australian wineries are well known for holding concerts with big international acts, while smaller wineries hold live music events where people come together to picnic and enjoy a tipple and tunes. If you have a natural amphitheatre, you might be able to host outdoor theatre productions put on by a local troupe, or show a movie on a big screen.

Mazes
People's love of solving puzzles means the maze never goes out of fashion. Mazes offer a chance for exploration, exercise and fun. A hay bale, corn or sunflower maze can be established in one season, but other mazes can take years to reach maturity. Plant mazes can be formed from all sorts of species. The Tangled Maze in Victoria highlights wisteria, jasmine and honeysuckle, the Bago Maze near Port Macquarie uses the

native species lilly pilly, and the Ashcombe Maze & Gardens on the Mornington Peninsula features everything from a lavender labyrinth to a hedge made up of more than 1 000 evergreen Cupressus macrocarpa bushes. Mazes can be grown from bamboo, roses and nearly any plant that holds its form and can be hedged as needed. Mazes take careful planning, extensive maintenance and a long-term approach to success. Seasonal mazes that change each year can be excellent for publicity and bring maze lovers back time and again.

Fossicking

The thrill of finding gold, a sapphire, a fossil or a thunder egg is a captivating experience. If your farm is in an area with a history of finds, you can offer this activity on your farm. But even if it's not, agritainment means you won't have to miss out on this visitor attracting activity. You can buy in the treasure, hide it and sell the opportunity to find it.

Art installations

Work with a local high school or artists to create exhibitions.

Simple attractions

Rather than one big thing, perhaps it will be adding lots of little things to your farm that creates a great day out for people. In addition to a farm animal activity you might also offer attractions like mini golf, water golf, croquet, quoits, giant chess sets, tyre swings, zip lines and obstacle courses, hay bale slippery slides, tractor tyre playgrounds, remote control cars and jumping pillows. You could use old fashioned water pumps, guttering and plastic ducks to create a duck-race attraction, or set up an area for an in-ground xylophone or pumpkin ten pin bowling.

Digging it

Who would have thought that some of the best rated tourist attractions in the world would be digger parks...giant adult sandpits where tourists get to drive excavators, bulldozers, backhoes and loaders. We're talking Diggerworld, not Disneyworld, and it might be worth you unearthing more statistics on the industry to see if a tourism attraction like this would work for you and your farm.

Museums and galleries
Perhaps you collect antique farming equipment, or have a taxidermy collection of unusual livestock or an impressive array of old farming books. Presented and marketed in the right way, a mini-museum on your farm might be of interest to key visitor groups.

Scavenger hunts and geocaching
Scavenger hunts have always been a fun way to entertain young and old, but now, with the advent of geocaching and hand-held GPS devices, it's been brought right up to date. Attractive to corporate groups for team building as well as to schools and families, geocaching is a clue to getting more people to your farm.

Paintball and laser tag
Is there a towering forest on the farm, as well as some outbuildings and structures? Do you mind large groups descending on your property planning to shoot each other? If so, aim for this target.

Zombie apocalypse
Is your farm spooky at night? Does the wind rustle through the fields in a *Children of the Corn* way? Are you good with set design and fake blood, and like the sound of chainsaws? If so, your farm could thrive on seasonal night time tourism where all you need to do is scare the living daylights out of visitors. You could decorate your biggest shed in a haunted theme, offer torches for people to go through a cobwebbed maze, or set up a full production incorporating a hay ride where visitors are accosted by fearsome sights, shrieking sounds and limping actors playing the undead. Scream tourism is not for everyone, but it can scare money right out of your visitors' wallets and into your bank account.

Agritainment is about entertaining people using your farm as a backdrop. It requires a lot of creativity and energy to get it right and it's more akin to running a theme park than an agricultural pursuit, but the two can combine in most unusual ways to help you make a life and a living on the land.

30

FUNCTIONS, CONFERENCES AND CORPORATES

Private Functions

Hens and bucks' do's

Have experience with hens clucking and bucks testing fences? Good, you'll need it when it comes to hosting the human versions. These strange creatures congregate for weekends away just prior to a wedding. Some groups are looking for adventure activities, some for pampering and some just for a place they can all hang out together and indulge in plenty of food and drink. Hosting hens and bucks' weekends isn't for everyone but it can be a niche to target, especially as these groups can fill gaps outside of the school holidays.

Parties and reunions

Everyone loves a party, and your farm could provide the venue. Birthdays, anniversaries, divorces, vow renewals and family reunions all need somewhere to come together and celebrate. Car and motorbike clubs are always on the lookout for somewhere to stop off for get togethers. You can choose what you want to provide. Perhaps you'll offer self-catering so they can BYO and just pay to use your lovely grassy area with access to toilets and a shelter in case of rain. Maybe you'll offer BBQ facilities or tend a spit or offer a full catering service. Work out the economics and what type of groups you need to target to make your farm venue a success.

Weddings

Whereas some couples choose a city skyline or beach background for their vows, there seems to be growing demand for rural weddings. Will weddings be a marriage made in heaven for your farm, or do you already know it's not the right business model to get hitched to? Providing a wedding-venue business on your farm will require impeccable customer service skills and attention to detail. Bridezillas aren't mythical creatures. They are real, and you don't want one on your farm unless you know how to deal with her, and her mother-in-law, and her drunken Uncle Larry, and all their long-lost friends. Even easygoing couples will make requests that will have you wondering why you didn't just stick to cauliflowers and cattle. Running your farm as a wedding venue can be a part time or full time business depending on your location and your offering. Some farms only offer their gardens and barn for the ceremony but not the reception, while some farms give the wedding party a blank canvas and it's up to the couple to bring in all the seats, tables and catering equipment. Others offer full-service weddings or something in between. It's your farm, so you get to decide how many weddings you will book in. It might be one a month or it might be several per day. Speak to wedding planners about what they require in a venue, research what other venues offer, and check out WedShed, where you can browse rural wedding venues online – and perhaps eventually list your own.

Corporate Uses

Meeting space

Do you have an area that can be used as a meeting space by local and visiting groups? It would need to be well lit (as well as able to be darkened when necessary for viewing presentations). It would need to be screened from insects, cushioned from loud noises and climatically comfortable all year round. A kettle for tea and a coffee machine, multiple power outlets and easily accessible bathrooms are other requirements for renting the area as a meeting space. This space could also be used to hold workshops.

Conferences and seminars

If you have a suitable meeting space, as outlined above, and there is also accommodation available either on your farm or in the surrounding district, you might be able to attract small to medium-sized conferences and seminars. Before you invest in facilities, speak with your local tourism organisation, as well as at least 5–10 professional conference organisers to gauge whether your region even needs a new venue, and if it does, what would help yours to stand out.

Team building exercises

Trust and cooperation exercises are all the rage for companies trying to bond their teams and get ever more productivity out of them. Sure, the HR team can do this in a setting in the city, but if you compare the vibe of an air conditioned office to that of the great outdoors, you can see there are opportunities to make fun and innovative team building exercises a niche offering on your farm. Speak with specialist corporate event and motivational organisers to see what they are looking for as break-out activities for delegates attending conferences nearby, or indeed as the focus of a day-long or multi-day team building event.

Corporate incentives

Workers in certain industries are offered a whole lot more benefits than just their pay cheque. HR departments employ incentive companies to manage schemes which reward workers each time they reach a performance or sales target. It might be a movie ticket, a bottle of wine, a dinner out, or it might be a ticket to a big sporting event, a weekend away in the country or an overseas holiday. Contact incentive companies to pitch how you could be a part of their program at different levels ranging from hampers overflowing with your value added produce to private farm tours or a weekend stay at your farm.

Product launches

Does your farm have a unique feature such as a twisting road up a magnificent mountainside, or a river that flows over large white boulders, or fields of flowers that captivate? How about century-old stables, a grove of olive trees or a ramshackle shed? If so, your farm might be great for corporate product launches ranging from cars to backpacks to romantic novels.

Location for shooting catalogues, commercials, TV shows and movies
Want to have bragging rights that Hugh Jackman, Margot Robbie and Jason Clarke (he actually did at our place) came to visit? Or that a bunch of models were strutting around your greenhouse manhandling your shovels? All those glossy magazine fashion shoots, TV ads, shows and movies need to be shot somewhere, and your farm and buildings might provide the ultimate backdrop. You can list your farm and buildings with private location scouts, or various government bodies such as Screen NSW's Reel-Scout database, which can be accessed by industry location professionals. If your property is eventually selected for filming, you will enter into a contract with the production company that covers everything from insurance and filming hours to the state they will leave the property. Some commercials can be shot in a day, but feature films can make use of a location for months. Not every farm is suitable for location shoots. It might need to be in a quiet area with no passing traffic noise, but funnily enough, needs to be accessible to large production trucks. Often, the closer you are to a major capital city the better your chances of being signed up, as crews and cast won't have to spend too long travelling.

The demands of tourists are completely different from the demands of hungry lambs and fungus infected leaves. Some people are adept at farming mushrooms, mangoes and mutton, and others are better at farming people. What might suit you best?

PART VI

31

FAMILY

LIVING, WORKING AND PLAYING TOGETHER

Y ou might be running the farm on your own and embracing the opportunity to run your own race, but it's also fairly common practice for partners and extended families to work together in small business. This brings with it additional challenges.

We've worked together, on and off, for about half of our 25-year marriage. We get along great, we wish the best for each other and we work at making it work. What's helped our relationship to stay strong over the years is that we have clearly defined skills, interests and work preferences. Whereas Andrew focusses on hospitality, livestock, machinery, accounting, building and general business, Anna concentrates on marketing, gardening, beekeeping, presenting to groups and making things like candles and balms. That division of interests means we're not always crossing paths and are free to make decisions in our chosen specialties.

Whatever your situation, hopefully some of the following ideas will help you get the most out of your living and working arrangements.

Allocate responsibilities

You know how it goes; tell everyone that everyone's responsible for everything and nothing gets done. But give people specific areas of responsibility and you'll have a better chance of success. This works for employees and family members who are employees.

You already know where your family's strengths lie, so play to them. Give each family member his or her own area of responsibility. It should suit their skills, personality and desires, and give them the freedom to make decisions that matter. It should also be appropriate to age and interest. Our elder daughter was great with horses, so she was in charge of horse care and guest pony rides. Our son preferred a different type of horsepower, so he got to do the ride-on lawnmowing and tractor work. Our younger daughter was the goat whisperer, and was responsible for helping with animal rounds and feeding.

If you're a couple, perhaps one of you will run the admin side of things while the other runs the agricultural side. Or one of you will run the greenhouses while the other looks after the open fields. There needs to be real clarity about the division of responsibilities so you don't tread on each other's toes.

Work talk

It's possible you'll still be talking about fencing, white cabbage moth and the rain gauge as you slip between the sheets. Don't. Set some rules about when you will talk about farming and the business and ban it at other times – including at the dinner table – so that not every conversation becomes a to-do list. We failed miserably at this and still survived, but it's not ideal.

Use technology such as email or WhatsApp chat to link members of your team so you can raise questions in a defined forum rather than always interrupting people with phone calls and texts.

Communication skills

When you do talk business, keep your personal relationship problems out of it and if one of you is working off-farm, keep communication open so both parties feel that their contribution is valued. Be a great listener, express yourself, but be kind, and take time to celebrate achievements, no matter how little.

Decision making

Consult widely with family members, but base business and farming decisions on research rather than whims and wants. Be prepared to explain why you made certain choices.

Conflict resolution

There will be conflict, and you'll need to handle it in a professional way. You can also bring in outside help to resolve issues. Don't be afraid to ask for help in doing so. If working with your children, keep everything fair so there are no claims of favouritism.

Change with the times

You don't want to be known as the old geezer who is stuck in the past. Stay fresh and up to date by being a lifelong learner with a positive attitude. Be open to suggestions rather than closed, and embrace change that is for the better rather than shutting the door on it.

Time away from the business

There's a lot to do when you're regenerating the land and growing food, and if you're combining it with off-farm work there's very little downtime. Exhaustion can set in through the perfect storm of big plans, big dreams, big days and way too big expectations of yourself.

"It's easy to stretch yourself too far," agreed Vikki Taylor of BurraBee Farm, Tuggeranong Valley, ACT. "By planning days off and having time for yourself, life just feels so much more manageable."

It's also about changing things up and not being wedded to how you do things.

"We wanted to take some time and slow things down for a while so decided one season to let Mother Nature dictate when we planted out the tomatoes," said Vikki. "Previously we've grown them in hoop houses, where we've worried throughout the season, *do they have enough airflow? Water? Are they getting too much water, is it too hot?* We'd roll up the side of the hoop house, *is it too cold?* We'd roll down the side of the hoop house, *is it too humid*? The worry – and extra work – went on and on, but when we decided just to let them be, the tomatoes were a little slower to mature but it didn't really matter in the scheme of things."

We know what she's talking about! The first five years we spent running the farm, we didn't need a holiday because it felt like we were on one long one. Everything was stimulating and fun, and it was like we were on a 365-day-a-year adventure. Then we hit a wall and needed a break, but couldn't work out a way to get away. We weren't willing to let the animals fend for themselves, ran too complex an enterprise to entrust it to a relief manager or neighbours, and so felt totally stuck. We finally worked out a deal with a repeat farmstay guest (he and his family had stayed with us 10 times), giving him a free holiday in return for taking care of the place. He won, we won (thanks Hayes)!

It's important to plan for breaks. Holidays help you appreciate what you have, rejuvenate, get fresh ideas and reconnect with your family without the ongoing chores and stresses that come with farm life, indeed any business for that matter. Plan well in advance to secure a farm-sitter or have staff trained up to take over the reins. Put the framework in place early so you never feel trapped.

Family comes first

The health and wellbeing of your family should take priority over the farm and business. Model great behaviour, make memories every day, encourage family legends and do things together that you'll be proud of when it's your turn to nourish the soil for the final time.

Exit plans and succession planning

What's your end goal? Perhaps it's just to enjoy the experience of farming for a while before you sell and retire. Or perhaps you want to build a business that will provide opportunities for your family for generations. Maybe you haven't even thought that far ahead! However, if there are children involved, you need to. Succession planning helps families work out the best and fairest outcomes in relation to inheritance expectations. There are specialised consultants who work with families to resolve these issues to get the best outcomes for all. It's best to start early.

Being able to work with loved ones is a true privilege that ensures lots of laughs, character-building moments and memories. Work out a way to make it work, and it will be an experience to cherish forever.

FARMER SPOTLIGHT: Bees

After 20 years working for customs and border protection across Australia's north, a son returns home to work alongside his father in a beekeeping business.

Farmer: Daryl Brenton
Farm name: The Beekeeper
Farm size: Home base is a 50 ha farm but the 1500 hives are transported around northern and north-western NSW
Average annual rainfall: Varies by site
Where: Home base is Kempsey, NSW
Farming since: 2004 (his father since 1960)

As his father began nearing retirement age, Daryl, who'd been working interstate for two decades, realised he'd regret it if he didn't spend a few years working alongside his father. He quit his job, returned to the family business, and it's now grown to employ an additional five people including his brother and nephews.

To prepare for his transition into beekeeping, he apprenticed himself to his experienced father. He then began mingling at key conferences, listening closely to the thoughts of other beekeepers and industry professionals.

Daryl believes the key traits needed by a beekeeper are patience, motivation, a calm temperament and organisational skills. Beekeepers also need to be able to cope with pain from multiple stings and to be prepared to work through exhaustion.

He spends a lot of time on the road, travelling up to eight hours from his home base. He needs to do this to check out potential beekeeping sites, meet with landowners and obtain permission to place his hives on their farms. He constantly needs to check weather reports to work out how nature will be affecting pollen and nectar flows in certain areas, whether roads will be impassable, and how dry spells might affect his bees in the months ahead.

Daryl says the best parts of his life as a migratory beekeeper involve seeing a whole lot of country across the seasons, seeing the bees at their best, developing relationships with property owners and being his own boss.

"I probably spend about seven weeks of the year camping out with the bees. The sunrises and sunsets can be amazing, but it can also be harsh if the weather turns or if even after all the planning, clearing of sites, obtaining permissions and travel, nature doesn't produce and the bees don't bring in a drop."

Working 60+ hours per week (95% on physical labour and 5% on admin/management), he sells his honey and beeswax via wholesale, retail and farmers' markets.

"To be a beekeeper, you need to be a real doer," says Daryl. "You need to be able to put up with long hours, lack of personal time and even on your days off you'll be thinking about bees, honey and all the things you've got to do. But it's literally a buzz of a business, especially when you discover even the smallest tweak that can make things easier and more efficient."

Food for thought from Daryl

What are the basics required for your farming activity?
You need a sweet tooth, bees, hives, extractors and access to flowering trees. You also need a licence from relevant regulatory authorities and your extraction area must meet health standards. If you want to get into beekeeping, work for a beekeeper first. It will open your eyes, and you'll be surprised how cheap honey suddenly seems.

What are the threats to the industry?
Pests and diseases such as American Foulbrood Disease, small hive

beetle and varroa mite. Cheap, inferior, adulterated and fake honey imports tarnish the whole industry. Oh, and bushfires.

Opportunities in the industry?
Increasing demand from Asia for pure, verifiable honey.

What makes a good farmer?
Someone who's a doer. You can't be lazy, you need to get stuck in and make it happen. You also need to be a problem-solver. It's great if you're handy and can fix vehicles and equipment. Physical strength in beekeeping is a must, and you need to be a planner so you can see problems and opportunities and have contingency plans that you're ready and willing to implement.

Best piece of advice from another farmer?
The only way to get something done is to start on it.

Worst piece of advice from another farmer?
"You won't get bogged if you drive over there." That's the last thing Daryl remembers hearing as his truck, fully loaded with bee hives, sank in mud up to its axles in a remote part of the country.

What would you farm if you weren't involved in your current activity?
I'm not really interested in anything other than bees, but maybe a cattle station in the NT just for the amazing country.

Daryl's top 5 tips
1. Really think about what you are doing.
2. When you buy second hand, you're usually buying someone else's problems.
3. Learn to enjoy what you do.
4. Keep good staff, get rid of bad ones.
5. Accept that no one will ever care or work as hard as you do.

What's the best way to enjoy your product?
A spoonful in a cup of tea!

32

WHERE TO FROM HERE?

O ur aim in writing this book was to share insights and information from small-scale farmers from around the country. We hope by doing so that you are better informed, and hopefully inspired, to create your own special life on the land.

A decade of farming has shaped our family, enriched the earth we worked above and hopefully enriched others. If we'd never farmed, we'd never have known the intense pleasure that comes from working with animals and nature and trying to work out how to break through our own limitations. We don't always get it right, we don't always get it wrong, but now we really do get it. Basically, for the sake of everyone and everything, Planet Earth needs people to connect with the land. Will that person be you?

Inspired? Scared? Want to know more? Rather than depositing resources here that will be outdated by the time you read this, head on over to the website at: www.smallfarmsuccess.com.au where you will find a whole host of free resources. Links to all the farmers mentioned within this book are available here and we'd welcome your feedback on what you'd wish was included in the book so we can write future articles for all to benefit from. Anna's also written a fun memoir, 'Honey Farm Dreaming', and you'll find more information about it on the website too. *Wishing you all the very best on your journey! Andrew & Anna.*

ACKNOWLEDGMENTS

To all the farmers who shared their stories in these pages, thank you for your generosity of spirit and information.

To Helen and Jack, Pat and Frank, we only made it to the end because of you. Big thanks for that, big hugs, and big heart for everything.

To Jill McIttrick, thanks for the good vibes and generous typo-hunting and to eagle-eyed editor Geoff Whyte (www.whyteink.com.au) – any remaining mistakes are ours not theirs.

To Nicolette van Straaten and families McCowan, Bedgood, Warriner, Tyrell, Kennedy, Trueman, Van der Meer, Grey, Martin, Keene, Moore, Tarrant, Hughes, Callender, Munnelly, Glimmerveen and Turnock-Jones thanks for coming with us on the ride.

To all our farmers' market friends, thanks so much for the laughs and encouragement. To Phoebe and Barry, knowing you were there over the years made everything easier. To our farmstay guests, Helpxer's, neighbours (Rob, Kerrie, Michelle, Ray, Phil, Jason, Kylie, Colleen and Rod), Tim and Kirri, Nabiacians and the fabulous Firefly Book Club, thanks for putting up with us too!

To Ian Wight-Wick for your encouragement when it was needed.

Finally, to our beautiful, resourceful, kind children, thanks for mucking in more than you ever mucked up. Big, never-ending love.

1

INDEX